Natallye Lopes Santos Oliveira

Jorge Amado's ontological geography

Natallye Lopes Santos Oliveira

Jorge Amado's ontological geography

A spatial dialogue with literature

ScienciaScripts

Imprint

Cover image: www.ingimage.com

This book is a translation from the original published under ISBN 978-620-2-40025-1.

Publisher:
Sciencia Scripts
is a trademark of
Dodo Books Indian Ocean Ltd. and OmniScriptum S.R.L publishing group

120 High Road, East Finchley, London, N2 9ED, United Kingdom
Str. Armeneasca 28/1, office 1, Chisinau MD-2012, Republic of Moldova, Europe
Printed at: see last page
ISBN: 978-620-8-31301-2

I dedicate this thesis to my late father, who tirelessly encouraged me to fight for life with determination and honesty, to appreciate classical music and to love Flamengo.

ACKNOWLEDGMENTS

First of all, Jorge Amado and the city of Salvador - where I experienced many andangas - for inspiring me to choose the thesis topic.

To my Bahian friends: Lua, Lipe, Joana, Luiz, Gabi, Roque and Juliana, who taught me a lot about being Bahian. Especially Nana and Drica, who, as well as being my beloved Bahians, also encouraged me to do this thesis with books, sofas and by sharing accents.

The University of São Paulo for the quality of its research program. The FFLCH postgraduate program, especially the Human Geography department, for encouraging intellectual development.

To my friends from São Paulo: Jana and Ale for sharing with me the moments of the disciplines at USP and many rounds of conversation.

To my mother, my great encourager, for dedicating her love, affection and energy to this long academic journey.

To my dear professor Dr. Ruy Moreira, for having been my advisor and friend for many years and for having helped me at the beginning of this process.

To Professor Dr. Douglas Santos, for having clarified many questions for me during the qualifying exam.

To my work and life friends: Rita, Raquel, Milena, Carol, Fabiola, Maria Alice, Alberto, Marcio Viveiros, Kelly, Isabel, Thais, Leandro Faber, Mariani Biteti and Vanessa, for their encouragement and collaboration.

To my lifelong friends: Gisela, Rafael, Manu, Leo, Luciano, Galdino, Aninha, Pris, Dani, Aline, Alinnie and Vivi, for always sharing moments with me.

To my mother-in-law, Fatima, for all her daily affection.

To my boyfriend Raphael, for his love and patience.

To my advisor, Prof. Dr. Elvio Rodrigues, for welcoming me, for helping me, for accepting my request and for introducing me to new issues in the field of geography.

To my companions and now great friends Felipe and Joao, for sharing all the anguish and joy of starting, building and completing a doctoral thesis.

SUMMARY:

OLIVEIRA, N. L. S. **Essay on an ontological experience in Jorge's Geography Amado**. 2016. 194 f. Thesis (PhD) - Faculty of Philosophy, Letters and Human Sciences, University of Sao Paulo, Sao Paulo, 2016.

This thesis was constructed from an interdisciplinary intersection between Geography and Literature. In a research proposal of an epistemological nature, this thesis proposes a definition of man in some of Jorge Amado's urban-themed works. Mar Morto (1936); Tenda dos Milagres (1969); Jubiaba (1935) and A morte e a morte de Quincas Berro D'Agua (1959) were the works chosen. We argue that this construction is only possible through a problematization of their geographies in a given geographical environment. Jorge Amado's works are all urban themed, considering a geographical reading of the city of Salvador with a concentration area corresponding to the downtown district and its surroundings. Themes such as: class struggle, racial clashes, miscegenation, precarious working conditions in the city, Afro-descendant culture and religion, female protagonism, popular social ascension, the power of the elements of nature and urban landscapes, were important subjects throughout the text. This thesis is based on an ontological reading, where man as a subject is the main theoretical concern. And the geographical foundation of man is based on his geographical environment. Thus, we conclude that the Amadian man to be defined is the one who is inserted in the city of Salvador, which also contributes to his conceptualization. The geographical environment of the city of Salvador in Jorge Amado's books is also an element that underpins man's being. Thus, Jorge Amado's geography is built on these categories of man defined and inserted in a geographical environment and bringing visibility to his geographies.

Key words: Geographicity, Geographic Environment, Jorge Amado, Literature, Salvador.

SUMMARY:

INTRODUCTION:

Geography was already part of the knowledge and experience of human beings long before it became a scientific knowledge linked to a disciplinary field. Geography has been a widespread field of knowledge since ancient times, with the whole Greek tradition of philosophical development. Geographical knowledge, or in other words, "doing geography", was born and developed from a social practice that existed long before the institutionalization of geographical science.

Since Geography is born out of social practices - and first and foremost, human practices - and considering that human practices are established from a set of experiences of/in the environment, it is possible to affirm that they are empirical, they arise from the real, the lived. Epistemologically organized knowledge with a geographical content rationalizes the empirical in question, also revealing the veracity of the relevant conceptual and methodological treatment. We are talking about the objectification of geography, the concreteness of geographical science, its realization. However, geographical knowledge is not exhausted by scientific systematization, since Geography exists in this relationship between man and the environment.

When considering new approaches to Geography, with the valorization of subjectivity in the relationship between man and nature, other knowledge began to be part of this epistemological construction: from Philosophy to Literature. The scientific paradigm became more complex and intertwined with different perspectives on reading the world. Thus, Geography is also gaining an increasingly dynamic field of openness to a transdisciplinary approach, or rather, to not limiting knowledge only to the dimension of its *episteme,* bringing together multiple axes of thought.

In order to continue this line of thinking about geographical knowledge, it is necessary to elucidate some of the reasoning behind this thesis. Geography or geography, as a foundation, is something that is established on the basis of the relationship between man/environment or society/nature. This is a well-established reading of what is determined as geographical knowledge. However, given the plurality of definitions that all these categories can have - man, environment, society, nature - Geography based on this relationship necessarily becomes plural in the same way.

By considering Geography as an ancient knowledge that predates the configuration of its science, it is possible to say that it is part of a world that was once much more guided by phenomenological observation as a method of knowledge than the current one of building knowledge under disciplinary logic. In this context, both philosophy and literature - the fields of knowledge of interest to this research - can be geographical because they deal with themes such as environment, space, place, history, location, distribution, nature, environment, relationships and man.

The theme of this research is to create a proposal for the construction of geographical knowledge based on the empirical treatment of Literature from an ontological perspective. In other words, by considering Geography as a legitimate field of knowledge since periods before its institutionalization, to develop a thesis that proves that it is possible to start from the assumptions of observing human existence/living in the environment in order to conceive a geographical knowledge determined by the experience of "being-in-the-world" having сото empirical basis in the field of literature.

The empirical content will be permeated by works by the Brazilian writer Jorge Amado (1912-2001). The novelist, a prolific teller of regional stories, came to be recognized as a writer who focused on the outcasts, fishermen and sailors of his land, who interested him as examples of attitudes vital to identifying a people and their environment. In the case of geographers, these profiles portray a social and spatio-temporal section that is fully satisfactory for observing the lived experience and existence of man relating to his environment from the real daily life repeatedly portrayed in his work.

The first part of the thesis, still part of the introductory part of the research, will present the epistemological construction listed in this objective of conceiving geographical knowledge from the ontological perspective of man and the geographical environment. In addition to these categories, we will work with the concept of geographicity, which is embedded in these two epistemological points of geographical science. Let's consider this part as the scientific foundation of what will be applied empirically from the literary content.

The first chapter will cover the *Dead Sea* book. After reading this text, we will look at the historical context of the publication, the author and the city of Salvador at the time. All of which are relevant to the initial stages of the thesis. Going into the book's narrative, we will realize that the name that gives the book its title is not a mere coincidence, since the sea is one of the main characters in the plot. The sea will be an ontological element in the class struggle, in the role of women in the plot, in the religious imagination and even in the connection with migrants. The main geographical issue is to make the sea visible as a fundamental element of the characters' being. Geography emerges from the sea and there will be no other way in this book.

The second chapter will look at *Tent of Miracles* as the book to be analyzed. The main theme of this book is the construction of the mixed-race man from Salvador and the rest of Brazil. The main character is also something unconventional, but in this case it is the very definition of the book's title: Tent of Miracles. This, which will be seen as the popular university of Bahia, is not a mere scenographic element of the geographical environment, but the space for the possible verification of this hybridity of the people that will be defended by Amado's miscegenated ideas. Tenda dos Milagres has the terreiro as an important element of its geographical environment.

The third chapter will analyze Amado's book *Jubiaba.* This is a book with a black and Marxist theme. Jubiaba is the character who gives his name to the title of the book, but he is not the protagonist, he is the father of saint of the terreiro of Morro do Capa Negro. Antonio Balduino is the main character and his movements from Salvador towards the interior will shape his geographical environment and the whole set of experiences of this character in the space-time of the plot will define what man he will become. Antonio Balduino is a black man who is born in Morro do Capa Negro in adverse conditions, experiences serious racial prejudices that determine his class and becomes, throughout his career, the black hero and leader of a trade union movement in the class struggle. The relationship between technology and man in his environment will serve as the theoretical basis for geography in *Jubiaba.*

The fourth and final chapter is a geographical reading of the novel *Death and the Death of Quincas Berro D'Agua.* In this book, Jorge Amado proposes a reading of the city of Salvador from a very subjective point of view that is presented to the reader through his protagonist: Quincas Berro D'Agua. Through this character's personal questions, Amado allows us to read the contradictions between the lower and middle classes in the city of Salvador. In addition, the settings for each of these classes portrayed in the book make it possible to problematize the geographical environment of the city centre of Salvador throughout the changes brought about by the "deaths" mentioned in the protagonist's plot. *Death and the Death of Quincas Berro D'Agua* was the work by Amado whose narrative was based on the identification and conflicts internalized in the archetype of an individual who can be perfectly seen as a common social being in the city of Salvador.

The central question of the thesis was to propose a definition of man (social being) in Jorge Amado's work, considering the section of the novels defined by the chapters. By analyzing the concept of man in the light of an empirical framework constructed by Northeastern regional literature, I was able to make its geographies perceptible throughout the research. In conclusion, we understand that these geographies are achieved through the determination of a geographical foundation for man.

The ontological experience in geography

Geography can be recognized as a field of knowledge in which one of its aims is to give us a way of conceiving things in the world. It dimensions the "being-there", the "being-af", the "being-in-itself" of the world. We can reflect that each conception of the world involves a type of logic, and a type of experience/existence is expressed as a result of this in the content of reality. Any knowledge of a geographical nature that is constructed necessarily involves man's existential condition. Therefore, since man is the founding element of the world, in order to understand the dynamics of both, the question arises: what is man? This is the beginning of an ontological debate. And considering man as part of the geographical dimension.

Reality would have a geographical foundation. Geography, or the geographical, as a foundation, is something that is established from the relationship between society and nature. This translates into a spatio-temporal order of the resulting elements. We believe it is possible to define this relationship as lived in the geographical environment. If we understand the specific scientific construction sought here as geographic consciousness, it involves understanding a geographic determination of existence.

Men organize their empirical lives according to the social relationships that exist in the environment, given in space-time. It is therefore in this context that existence takes place, because the relationships announced there lead to the movement of being and non-being. And that's what existence is all about: being is always ceasing to be (Martins, 2007).

So how can this relationship between geography and ontology be worked out? In an attempt to define what the geographical element of reality is. Since Ontology is an investigation into the need to define things as they are, especially the essential contemporary question of what man is, we will therefore deal with things that exist, in other words, reality. The geographical foundation is what the ontological debate in Geography is about. It is present in reality. If we limit *geography* to what geographical science produces, we lose its importance and significance in the constitution of reality.

The big question that follows is how to identify this geographical foundation in the reality that surrounds us, that we build ourselves, that grounds us and recreates us ubiquitously as we live. Identifying the geographical foundation of reality means clearly perceiving geographical phenomena. It's as if we were looking at the world and asking ourselves what is geographical about it (Martins, 2007).

A relevant consideration regarding the ontological reading of Geography defined under the bias of the foundation of reality comes from Correa da Silva (1978):

> "[...] insofar as Geography's true 'laboratory' is reality itself, its field encompasses the whole complex network of human and natural phenomena, as they present themselves to observation, in the given environmental conditions, which cannot always be reproduced under conditions of controlled experimentation [...]" (page 75).

Correa da Silva's statement above that Geography is a field of analysis of reality through observation of the world, allows us to say that this knowledge is much earlier than any scientific systematization. This is because, through the senses, the world is already susceptible to our attention and possibilities of description. Geography is a foundation of life, geography is read in everyday life, in the changing relationships between man and the environment.

The apprehension of the reality of the world around us, via the senses and the description of the environment that is observed, bringing to light the geographic foundation, can be read as the birth of human consciousness. Being conscious is knowing how to (re)know this reality of the world that surrounds us. Consciousness is the result of all this reflective objectification of reality and our ability to grasp it. We agree with this reasoning:

> "Our existence, and the awareness of this existence as men, occurs to the extent that we designate, conceptualize the reality that surrounds us, that is, our otherness, the environment" (Martins, 2007, p. 14).

Correlating these ideas, it is now possible to affirm that our understanding of the world, our comprehension of reality, is based on knowledge of the existence of beings. In turn, this existence takes place in the world, and is the world. Thus, it is in the spatio-temporal reality of beings that our understanding/construction of the existence of the world emerges. Since Geography is in the world, in the factual reality of things and beings, it is now a category of existence. The question arises as to what this Geography is, with its foundations in reality. If we don't reflect on this, Geography will be limited to the perimeters of Geographical Science, and the fact is that it is beyond that.

Geography, in an ontological reading, as a category of existence, will nevertheless provide a scientific framework for research. From geographical science we obtain a certain representation, in thought, of the objective reality that surrounds us. Thus, through its methodological procedures, science structures itself in different interpretative theories, building a conceptual framework and defining its main categories. Geographical science, therefore, is an act of theory in practice, of theoretical practice, and at the end of the day it even represents its institutionalization (Martins, 2007).

The geographical categories of ontological analysis

Science has a political and even subjective purpose because it also responds to circumstances resulting from a historical and social context. Through its methodological procedures, science is structured around different interpretative theories, building a conceptual framework and defining its main categories. Analyzing the current period in the history of science and its different trends and matrices, many argue that there is a crisis in the meaning of producing knowledge today. With the so-called *Post-Modernity,* we are experiencing a moment in contemporary history when culture and knowledge are fragmented. It is, to a large extent, the story of a change in the way we view the unity of knowledge. Identified more and more as "specialized" and less and less by the "holistic" production of knowledge, questions about the validity of the way of doing science have been consecrated and - especially positivist and Cartesian - have multiplied.[1]

A very important aspect that cannot be evaded under any circumstances is to bring visibility to those who are the protagonists of scientific endeavor: man. Science is thought up by man, planned by man, executed by man. However, it is less and less for man, in a subjective sense. Under the aegis of this reflection, the ontological debate becomes increasingly relevant, because it is necessary to say which scientific man is the one who produces knowledge and whom this science, as it has been developed, serves. What are the objectivities embedded in scientific and geographical discourse and which subjectivities do they represent and serve?

How has geography been practiced within this scientific context? Evidently, from geographical science we obtain a certain representation, in thought, of the objective reality that surrounds us. This is a process of subjectivation that goes through methodological procedures. A subjective representation established through logical systematization, expressed in one or more languages. All of this points to the phenomenal and essential aspects of reality. With the epistemological sense of existence that will be defended here throughout the text for Geography, and a reading of the world, it

[1] The adjectives "positivist" and "Cartesian", used to associate scientific practices, come from an established period of Modernity in which scientific production was seen as the truthful answer to what the world is.

is now necessary to identify the theoretical and categorical framework that will be used to carry out the thesis. These are the categories that will fill the discourse with meaning: *geographical environment and geographicity.* The geographical environment will be used as the theoretical resource for understanding the spatio-temporal nexus of the world, where the social relations of existence are constructed. Geographicality, the nomenclature of the geographical foundation of being, will be the concept that gives greater concreteness to the ontological discourse in Geography.

A final reflection on the choice of categories that will be applied theoretically in the research proposed in this thesis is the question of their importance in Geography. Or rather, why these chosen categories consistently fit into a geographic scientific production. The argument here will be that both consist of making explicit the relationship between society and nature. Or, man and environment. Other categories also considered would be work and gender. However, these reflections and applications will be discussed at a later stage in the research.

The object of study of Geography within this ontological reasoning for the geographic foundation will carry the status of *geography.* This emerges from the relationship between man and nature. And it is from the dynamics of this relationship that we observe the being of man. This "object" that possesses geography emerges from the co-existence and alterity of man in relation to the environment, of society in relation to nature. In a brief definition of the understanding of Geography, we agree with the following conceptualization:

"Geography or geography, as a foundation, is something that is established on the basis of the relationship between society and nature. In other words, in both subject and object, the relationship between society and nature translates into a spatio-temporal order of the elements that result from the relationship. To put it more bluntly: when the relationship occurs, a fundamental determination of existence takes on meaning in act and potency." (Martins, 2009, p. 4)

Geography is therefore born out of the relationship between society and nature. It is the property of man and the property of the environment. But it is above all the property of a whole, in which the environment and nature are the extension/projection of man and society insofar as they are the creation and representation of the geographical foundation of existence.

1. The Geographical Environment

The simplest and most general way of defining the geographical environment, considering that it is the result of the relationship between society and nature, would be to define it based on the intersection between the natural environment and the cultural environment. The natural environment and the cultural environment form the geographical environment. This changes over time as natural history and human history develop. And by analyzing its dynamics, as the elements of the geographical environment are related to each other, the natural or cultural links that unite them define a space of relationships. A detailed definition of the geographical environment is as follows:

"[...] when society appropriates nature, it imprints an order on this objectivity, which is expressed by geographical principles. And appropriated nature becomes a *geographical environment, from* which point the relationship becomes society/geographical environment. In fact, the process of subjectivation/objectivation in the construction of the geographical environment takes place through geographical principles as a dimension of existence, both for the subject and the object." (Martins, 2009, p. 5)

The geographical environment can be translated as the order imprinted on nature through man's sociability, in which he constructs his geography and dialectically reconstructs himself as well. Where does geography fit into this proposition? In an existential reading of it, what exists is the environment. Insofar as it is consciousness for man. And because it exists, it has geography; something that emerged from the relationship between man and environment. Thus, the best way to grasp the meaning of the geographical environment is to always try to interpret it in relation to man.

The relationship between man and the environment also serves to designate geography. And in order to sew this relationship with the ontological basis of the geographer, it is enough to understand the reflection:

"As an extension of himself, of man, in the double sense of appropriation, by designating the Geography of the environment, the objective bases of the Geography of Man are established. And the Geography of Man, as the ontological foundation of his being, is established as subjectivity." (Martins, 2009 page 24)

When analyzing the properties described about the understanding of the geographical environment, a certain degree of caution in interpretation is necessary. The hasty idea that the constitution of the geographical environment is a uniquely harmonious process in a state of total equilibrium would be a misinterpretation of this category. The contrasts embedded in the spatialities of the geographical environment are not only present and make up the reality that forms it, but are also necessary in view of the otherness that exists to identify the world. Among geographers, the different readings of this concept allow for a broader understanding of this otherness. In a phenomenological reading of the geographical environment, considering both natural and social aspects, Dardel (2011) points out differentiated spatialities:

"In all peoples there are two types of spaces, revealing two different cosmologies: a surrounding world in which human presence and work are manifest, populated regions, cultivated lands, navigable rivers, frequented seas, accessible mountains; a disturbing world, deserted expanse, wild land, unknown sea where no one has penetrated." (page 62)

In a more critical and Marxist reading of geography, the geographical environment can have another connotation:

"In geography, the relationship between man and the environment is a movement understood as the search to satisfy the material needs of subsistence, which is also the end of its realization. In Marxism, it is a process that begins at the level of awareness of the material condition of human existence and culminates in the realization of man, his hominization, as a fully realized being." (Moreira, 2004, p. 29/30)

Man now emerges within the relationship between man and nature and then enhances the geographical environment. Survival and existence require a relationship between man and the environment, between society and nature, in order for a choice to be made, and for this choice to be defined in work. Work, another essential secondary category in this thesis' reasoning, and the basis for understanding the geographical environment, will be discussed later. It is known that a concept such as geographical environment is not exhausted and limited to the reading presented here. A more in-depth problematization of it is important and this gap is acknowledged here.

2. Geography

In order to discuss the conceptualization of the category of *geography,* three different but not necessarily divergent views will be addressed: Martins (2007), Moreira (2004) and Dardel (2011). And, with regard to Dardel's thinking, considerations based on the view of Besse (2011).

What is this geographical foundation, this proposal for an ontological reading of the world called geography? Initially, geography will be present in the reality constructed through the relationship between man and the geographical environment. Furthermore, from now on we can also designate geography from the point of view of geographicity, since it becomes the existential foundation that gives rise to the adjectives that constitute being. The ultimate expression of man's being is concrete through the materiality of geography.

Martins (2007) attaches great importance to geography. He even states throughout his text, as he does at various times, that geography is understood as a synonym for geography. This, in turn, becomes tangible and its understanding is achieved through its description as a possibility. To this end, this description carries attributes and properties. It comes from:

"observation of the phenomenal present in the materiality of the world that surrounds the subject.[...] And for the description to be realized, as an expository act, it has to have the sense of cohabitation and co-belonging. [...] In other words, decoding requires a cognitive procedure in which, beforehand, things need to be perceived in their location and, consequently, in their distribution. This is what the sense of the geographical invokes in its most basic constitution." (Martins, 2007, p. 16)

The sense of the geographical and the reading of geography. An important consideration according to this reasoning for understanding geography is that it is most easily achieved through a sense of location. In other words, in Geography, the essential question for reading the geographical environment from an existential point of view is "where?". Thus, geography is determined according to the logic of location, of being-there in the world.

From another methodological perspective, this is the reasoning of Dardel (2011). In this definition, the subjective content of the analysis of reality and the scope of geography is much more emphasized. Geographicity is like that:

"Love of the native soil or search for new environments, a concrete relationship links man to the Earth, a geography of man as the mode of his existence and his destiny." (Dardel, 2011, Page 1)

In this phenomenological/ontological reading of geography, another important consideration is that land and man are not merely a complementary and defining pair of reality. Here, in an objective-subjective relationship, they will be read as an inseparable pair. One is founded and recreated in the other. They are conditioning existential elements. They cannot be thought of geographically separately. Problematizing Dardel's geography, Besse (2011) states:

"Dardel links landscape to what he calls human 'geography'. The choice of this term is not gratuitous. It signifies the insertion of the earthly element among the fundamental dimensions of human existence [...] It is necessary, he reminds us, 'for man to feel and know himself to be connected to the Earth coтo to be called to realize himself in his earthly condition'." (page 120)

Dardel's reasoning, in a Bessian interpretation, opens up a new ontological perspective for geography. Man, in this defining nexus of geography, has a whole meaning defined in a very particular way. In this preamble, the scientific crisis is criticized and overcome: "[...] in addition to claiming to overcome metaphysics on the basis of logic and technology, he advocated the autonomy of science from philosophy. Dardel unites science and philosophy to perfection." (Holzer, 2011. Pag:148)

In a third conception of geography, its foundations are defined in a spatial logic of reading geography. Moreira (2004) uses the expression "the being-of-man-in-the-world" to refer to geography, an expression that will be repeated in various other publications by the geographer.[2]

Moreira will also read geography from the point of view of man's being. According to his reading, geography is in itself the synthesis of the relationship between the concrete essence and existence of being. He designates:

"Geographicality is the spatial condition of man's existence in any society. The equivalent of what in philosophy Heidegger calls the worldliness of man. Or, in another context, Hegel refers to man's being-in-the-world." (Moreira, 2004, p. 32)

---

[2] Moreira (2004) , Moreira (2010).

[3] In Geography, the concept of place carries an epistemological sense of belonging to a given spatial area for those who identify with it. In this case, the sea is a place for sailors. Theorists of this concept include: Sauer, Carl (1998 [1925]); Holzer, Werther (1999).

Once again, we would like to emphasize that the concept of geography does not end with these considerations. Much less has all the problematization that will be put forward as pertinent to the thesis been finalized. It was up to the text to highlight the importance of the category for the desired theoretical arrangement, the geographical meaning that can be achieved through the use of this category and the further development of the thesis' ontological proposal.

Man *and his definition for Geography:*

Carrying out research with the proposal of an ontological debate for scientific construction, at some point, involves defining what man is that we are talking about and problematizing. Since Ontology is a field of reflection on man, its objectifications/subjectifications correspond to the problematics of this context. The construction of the entire theoretical framework developed so far moves towards a critique of the established practice of science: the increasingly abstract content of man and his subjectivity in the construction of this respective knowledge. So, since one of the most important considerations of the research is to ask what kind of man is sought in an ontological reading, a brief attempt at a definition is in order. The following is a preliminary proposal as to which man - theoretically speaking - will be sought as a result of this thesis.

The first important consideration comes from the unique nature of man in relation to other existing beings. In addition to being endowed with reason, from this perspective, Martins (2007) emphasizes:

> "Man is unique in that he is the only one capable of being both subject and object. He is the subject who objectifies himself. This is because man, in view of his capacity for rationality, is the only being capable of inquiring into the definition of himself and his being" (page 35).

This is a cognitive reading of man. It does not meet the biological or physical conceptualization of man, for example. In line with this previous logic, man is the only being capable of being both subject and object of himself ubiquitously. Therefore, what exactly singularizes us as a species is this ontological foundation, which together with our biological universality places us as a particular reality.

The aim of discussing and defining the pre-conceived concept of man for this thesis is to provide a philosophical systematization of this idea of man with a view to establishing an ontology of the human being capable of answering the classic problem: "what is man?". Reductionism in understanding what man is is then identified as science's biggest problem. With this, the role of Ontology grows in relevance. Especially since ontology is the part of philosophy that reflects on and theorizes about the fundamental nature of man, the "being-of-man".

The manifestation of these reductionisms is via epistemological propositions, and falls under what we are identifying as philosophical problems in the field of human sciences. Like geography. Among these reductionisms, Lima Vaz (1992) states:

"With the enormous development of the social sciences, the economic and social spheres were incorporated into the ontological debate, linked to the concept of work as production. Economic (Marxian) and sociological (Durkheimian) reductionism emerged from the discourse on the essence of man as a social being and producer. [...] Another reductionism is related to the problem of History. The historicity of man becomes an important theme in ontological reflection." (page 11)

Let's add to these the problems of geographical science: the dispute between geographical determinism and the recognition of the geographical foundation (generator of geography) as a founding element of the being-of-man. Thus, in this vast horizon of the sciences of man, old and new problems present themselves that will constitute, together with the permanent data of natural experience, the domain of man's knowledge about himself that philosophical reflection will have to systematically thematize and organize around the ultimate center of man's intelligibility, which is his self-positioning as a subject." (Lima Vaz, 1992, p. 13)

We are aware of the plurality of meanings of the definition of man and its founding elements that construct an ontological perspective of knowledge. Therefore, it should be emphasized that the approaches that will be made within this proposal are not exhaustive, but are strategically chosen to conceive the concept of man that will be problematized throughout the thesis. In other words, these ontological problematizations that will now be focused on have guided the ontological sense of man that will be defended throughout this thesis.

The ontological experience can be problematized and encompassed by different conceptual proposals as well as many other theories not mentioned here. However, our main research objective is not to list the possible theories pertinent to the ontological debate in Geography - even though we believe that we have not exhausted the possibilities with this theoretical survey. In this thesis, the main issue will be to carry out a reading of the ontological experience as theorized in a given section of Jorge Amado's work. To turn ontology into a political field for Geography as an enriching theory on the definition of man and the spatialities that surround him. The way in which Jorge Amado developed not only the multiple existences of his characters - real people from the urban Bahia of Salvador - but also their environments, didactically contributes to the propositions raised in this thesis.

CHAPTER 1 - "DEAD SEA" ON STAGE

1- Jorge Amado and the Dead Sea: the spatial context

This thesis will explore how the city of Salvador or some of its features are relevant in understanding who the man who lives there is. The books to be analyzed are works by the Bahian writer Jorge Amado. To follow the trajectory of this proposal, some of his novels in which this perspective was seen as appropriate will be listed.

Amado was chosen because he was a writer who repeatedly produced novels that were much more denunciations of a reality than mere aesthetic objects. Undoubtedly, the city is one of the places where the performances of everyday life in Amado's works take place. It is through the city that we can perceive the (re)production of society in the making of its history. Within this framework, it is important to say that there is no single image of the city. There are various images that construct an image of the city. It is through the various processes of constructing this image of the city that the imaginary about it emerges. The imaginary is the set of elements that make up the essence of a city. In this specific case, the main landscape of our object of study. Therefore, loaded with subjective content as well, the city imaginary is not presented in a single form, as it will be defined based on the particularities of those who seek to apprehend it.

In the problematization of this thesis, we call geography the relationship between man and the environment through an ontological experience of existence, now also seen as the way in which we encounter and see this imaginary. Where the lived fabric of being is elaborated in the space-time dimension of a specific location. It is believed here that literature is perhaps the purest form of apprehending geography.

In order to investigate the imagery and geography contained therein, or even more directly, the images that the city is capable of providing and producing for our senses, it is important to understand the historical dimension in which it is inserted. Because of this, it is important to stress that the aim of this analysis is not to question or address the contradictions - which do exist - between the author's approaches to the city and what the present imposes on us as truth. Nor is it to discuss whether or not there are ideological contradictions in Amado's life that influenced his approach to the city, hiding or highlighting facts. The aim is to grasp what is geographic in the discourse of his works, which will be discussed here.

Jorge Amado (1912-2001) was one of the most prolific and recognized Brazilian writers both here - on a national scale - and abroad. Studies of the novelist's work can be divided into different aspects, such as gender, race, politics, religion and national identity. However, these are admittedly important points, but in this thesis, they will circulate as a backdrop to our analysis. The important thing will be to look for and identify how elements of his geography are addressed throughout the novels and are undoubtedly points for understanding the human figure that Amado created throughout his work. The being of the Amadoan man is built and founded on this geography.

Dead Sea (1936) will be the first work to be discussed in this first chapter. There is an interesting homogeneity within the writer's vast literary heterogeneity. The sea, one of the most significant geographical elements of the city of Salvador, is the central character of the work, even if this is not emphasized so clearly throughout the text. Throughout the narrative, the sea was one of the city's gifts. It was one of the writer's most important sources of inspiration. It is the setting, the main character, a mythological link, a means of carrying out religious practices, and nature in relation to man. Throughout the narrative, he manages, from various points of view, to express feelings of admiration, fear, respect and passion for this enchantment that nature has reserved and given as a gift for the existence of the city of Salvador.

One fact is indisputable in Jorge Amado's novel *Dead Sea.* We can see that the author does not approach the sea as a pure and simple natural element. He makes a perfect connection between the sea and the people who enjoy it. Often, this natural element is reified, gaining an almost life of its own, even if it is through the mythological/religious figure of Lemanja/Janaina and all the other names that recognize it in the novel. The sea doesn't just appear as part of the landscape, making up the scenery. In this plot that reveals the Amadian geography of the novel, our geographical imaginary, the sea is an "actor" on stage, giving life to the environment that cohabits with society. From there, our geography emerges. Thus, we will take from the text this living empiria, this context in which the ontological experience in the work emerges as a possibility for analysis. We will engage in a dialog between the geography of the novelist and the geography of the geographer to help build the geographic foundation of the thesis.

2 - *Jorge Amado and his Salvador in "Dead Sea"*

Salvador, more than any other Brazilian city, carries within it the country's past and present. In this thesis, we don't just want the "city-habitat", but the city that is inhabited, that has habits, what in popular language is called: people and not numbers and their statistics. The city of man and not of population concepts. The city of human relations and not merely of economic activities. I believe that Jorge Amado has a prominent place in the attempt to portray this more humanized city. According to geographer Auxiliadora da Silva (2004), "no one has told us better than him about the sea, the alleys and lanes, the steep streets and the mysteries and magic of the city of Salvador" (page 27).

Any doubts that the city is a source of inspiration for writers and a daily challenge for its inhabitants, as proposals for a geographical humanization of Salvador, will be resolved throughout this thesis. This is why we believe that it is important to look at the novelistic work of Jorge Amado, intimately with the city of Salvador.

Amado lived in Salvador both as a youth and as an adult. His novels tend to represent snippets of the city from its central neighborhoods, as they were the most familiar to the author.

Salvador was initially born to be a "strong city": it was an urban space with an important military strategy. As this urbanization progressed, other relationships were born and developed there. However, it is important to realize the importance of the relationship with the sea in the very foundation of the city.

[a]Written in the neighborhood of Gamboa de Cima, in Salvador, Bahia, on the seafront, and completed in Rio de Janeiro in June 1936, the novel was awarded the Graga Aranha Prize in 1936, when its first edition was published by Livraria Jose Olympio Editora, Rio de Janeiro.

Mar Morto (Dead Sea) is the story of Guma, a child who grew up on the docks of Bahia, and his love for Livia, who, after the death of her beloved, takes his place as a saveirista, gliding through the depths of the sea from, literally and metaphorically, her safe harbor, Bahia de Todos os Santos, in the city of Salvador. In the very opening of the book, Jorge Amado states that he came to tell the stories of the men on the edge of the docks of Bahia in these pages. He also calls them the "people of Lemanja", building up the importance of the role of the sea, a living character in the book, and of paramount importance in understanding the narrative.

One of the thematic universes that is very well portrayed in the novel *Dead Sea* is the clash between land and sea. The author has created a dualism that firmly portrays the reality of the clash of classes between the sailors/shippers and the employers/traders. It's a clash that's at times turbulent and at times silent, but the respective scenarios of their existence - sea and land - essentially define which man is naturalized in each environment, retracing the full history of each one.

3 - *The sea at the Dead Sea*

There are different approaches to this character/scenario in the novel: the sea. The sea is the one that, in connection with the different phenomena of nature, either embraces or swallows its men. The sea as a phenomenon of nature and the very grandeur of its protagonism throughout the book. The sea is the sailors' shelter. It embraces them in the solitude of their work far from home. The sea is more the ground of the sailors' identity than their own cities. The sea is the "place"[3] in a conceptual geographical sense of the sailors. The sea is what gives sailors freedom. Freedom that doesn't come from work, money or adventure. Freedom comes from this intertwined relationship between man and sea in the Amadian work analyzed here.

The writer opens the book in his preface by saying that he came to tell the stories of the men on the edge of the pier. According to Amado (1936):

> "The old sailors who mend sails, the sloop masters, the tattooed blacks, the malandros, know these stories and these songs. I've heard them on moonlit nights on the market quay, at the fairs, in the small ports of the Reconcavo, next to the huge Swedish ships on the bridges of Ilheus. O people of lemanja have a lot to tell."

The opening of the book already highlights the importance of the sea for the continuity of the novel. A storm hits the area frequented by the schooners and devastates the sailors, causing fatalities. Livia was distressed, on the edge of the quay, in the rain and wind, waiting for Guma who was coming in the Valente (the fisherman's boat), defying the fury of the winds. A sloop capsized in the sea and two men (Raimundo and Jacques) walked into the water and died. Rain, wind and sand talk to the sea, and in this rhythmic relationship, the cycles of nature take on a mystical air. The sea is grand and alive. "The rain came with fury and washed the quay, kneaded the sand, rocked the ships at berth, revolted the elements, made all those waiting for the arrival of the ocean liner flee." (Amado, 1936 page: 18)

The plot of the characters begins to be presented to the reader from this event around the grandeur of the sea. It will therefore be the element of coexistence between the people who will develop their relationships throughout the book. The man, the Bahian born in Salvador after the *Dead Sea,* is the one who maintains this intrinsic relationship with the sea. Man's being is defined by the sea. Two passages are interesting about this coexistence:

"The men on the edge of the pier only have one road in their lives: the road to the sea. This is the road they enter, and this is their destiny. The sea is the master of them all... The sea is unstable. As is the life of the men on the sloops. Which of them has had a life like that of the men from the land who cherish their grandchildren and bring their families together for lunches and dinners?" (Amado, 1936, p. 25)

"What if, one night, they came to her with the news that Guma was at the bottom of the sea and the Valiant was wandering aimlessly, rudderless, without a guide? It was only then that she felt all of Judith's pain, that she felt totally her sister, sister also to Maria Clara, to all the women of the sea, women with the same fate: to wait out a stormy night for news of a man's death." (Amado, 1936, p. 24)

The first passage portrays quite clearly how the sea is an existential element of man's condition in the novel. The sea identifies him, individualizes him, differentiates him from the men of the land, who were not born on the edge of the pier, or are only different from those who do not develop working relationships through this setting. Another important reflection here is that the sailor, the Amadian man of the *Dead Sea*, is essentially defined by his working relationships. To be a sailor is to be identified from a working condition. Therefore, work is implied in our reflection - as mentioned in a previous theoretical moment - as the activity that also defines us. Who would a sailor be without the sea?

The second passage is a reflection from the perspective of Livia, the novel's romantic heroine, who wasn't born on the edge of the pier, but became absorbed by it on a daily basis, to the point where it also became a defining element of her being. Livia fears for the life of Guma, her love, on the voyages of her sloop Valente on stormy days. Once the storm has passed, Livia continues to wait for Guma and hears Maria Clara moaning inside the sloop with Mestre Manuel. Rufino tells Livia that Raimundo and Jacques drowned and their bodies were found by Guma. Everyone shares the suffering of Judith, Jacques' wife, a mulatto woman who has been left with a child in her belly. Maria Clara is still sobbing with love. Judith won't have love tonight or ever again, because her man died at sea. From the abandoned fort comes the music sung by the old soldier Jeremias, the powerful voice of a black man:

"The night is for love...

Come and love in the waters where the moon shines...

And sweet to die at sea... " (Amado, 1936)

Amado's use of the sea in his characters is remarkable. The natural element is an intrinsic part of the lives of its inhabitants. In various chapters throughout the book, Amado reveals the importance of the sea in the lives of fishermen, sailors, their wives and girlfriends, in short, all those whose lives are connected to it in one way or another, including himself - the writer of that period.

Parallel to this existential relationship between men and the sea in the novel, Amado also tries to make this journey, with great romanticism, into traces that are part of the daily life of Bahian culture, in these natural elements that make up the landscape. It's a dialectical relationship, because at the same time as he uses these components to elaborate the proposal of his writing, an imaginary of the city seen from the deck of the sloop, from the edge of the pier, is perceived by those who don't know it. It is possible to get to know the city of Salvador from the point of view of these sailors. In this way, you can't define who influenced whom. Whether it was the city that gave rise to Amado's imagination, or whether it was he who, through his vast published works, made us perceive the city in his own way.

It is also important to emphasize that the totality that makes up the city of Salvador as a place born from the cutout of the edge of the pier, seen from the point of view of those who sail, of those who are inserted in the sea, very much perceived in this work by Amado, is a fragmented view of reality. It's a partial view of the city, the sea and the creature itself, established through the existential logic of the man we work with here. The geography we read in Amado and his "Dead Sea" comes from his perspectives, from his experience, cutting out what was significant to him. It is a romantic vision, undoubtedly nostalgic, of a mythical and complex city, but also strongly unequal.

4 - The mythical sea of Lemanja

Dead Sea is a book filled with an imaginary of social realities. However, in addition to this, another theme is developed throughout the novel which also reflects the importance of the sea in people's lives: the religious element. The sea in *Dead Sea is* the sea of Lemanja, Janaina, among other names given to the goddess who inhabits this immensity. This entity will have an incalculable value in the sailors' lives through the religious faith and messianic tone in which she appears throughout the text.

Religious syncretism is another Bahian trait that is often portrayed throughout Jorge Amado's books. In *Dead Sea*, this trait is also present throughout the text. This syncretism is represented through the different conceptions of lemanja, the religious entity of the sea. From a spiritual entity of Afro-descendant religions to a representative of indigenous legends. There are countless facets associated with the queen of the sea. Take a look:

"lemanja, who owns the quays, the sloops, all their lives, has five names, five sweet names that everyone knows. She's called Lemanja, she's always been called that and that's her real name, mistress of the waters, mistress of the oceans. However, the canoeists love to call her Dona Janaina, and the blacks, who are her most beloved children, who dance for her and fear her most of all, call her Inae, with devotion, or make their supplications to the Princess of Aioca, queen of those mysterious lands that hide on the blue line that separates them from the other lands. However, the women of the pier, who are simple and brave, Rosa Palmeirao, the women of life, the married women, the mothers who are waiting for fiancés, call her Dona Maria, which is a beautiful name, even the most beautiful of all, the most revered, and so they give it to her as a gift, as if they were taking a box of soaps to her rock on the Dike. She is a mermaid, the mother of water, the mistress of the sea, Lemanja, Dona Janaina, Dona Maria, Inae, Princess of Aioca. She dominates these seas, she loves the moon, which she comes to see on cloudless nights, she loves the songs of the blacks." (Amado, 1936 page 78, 2p)

lemanja is the entity of four names. The people of the sea are the subjects of Queen Janaina, Lemanja. Their identity is defined by their relationship with the sea. Men who are of the sea do not go to the land to work in any other profession. The spell of this lioness constructed by Amado is very strong. The women of the pier treat it with respect because they know that all the men who live on the pier, their husbands and/or relatives - and in the future their children - embrace the destiny given to them by the sea. In other words, in their beliefs, fate is woven by Lemanja. For seafarers, she is the mother of water, the mistress of the sea, and all the men who live on the waves fear and love her. According to their belief, she punishes because she never shows herself to men except when they die at sea. For this reason, she is loved and feared. About Lemanja, and the relationship between men and women:

"Then the moon came and Janaina's hair spread out over the sea. Then music came from the sloops, from the old fort, from the canoes, from the pier, greeting the mother of water, the mistress of the sea, whom everyone feared and desired. She was mother and wife. Only she knew their desires and only she consoled everyone. The women now prayed to Lemanja." (Amado, 1936, page 42)

The fate of the sailors was a designation derived from the wishes of Lemanja. They believed that their financial prosperity, the woman they loved, contact with God and death itself were designs offered by her. From this, we can see that the Amadian man is a man of faith, practiced from the elements of nature. The empirical, materialism experienced by the sailors also comes from faith in the religious imagination.

Through the mythical figure of Lemanja, then, the sea takes on a supernatural tone in Amado's work. As a defining aspect of man, the sea is his dwelling place and also his space of connection with God.

5 - *The duality of land (city) and sea (pier)*

The landscape that permeates the entire novel *Dead Sea* is dualistic: it is portrayed from this comparative clash between land and sea. This duality is emphasized as a consistent element of reference to the geographical environment. Guma, the main character, the sea-hero of the book, is constantly identified in the novel through his memories. Note the following passage:

"On one side, huge and illuminated by a thousand electric lights, was the city. It climbed up the mountain and its bells tolled, from it came joyful music, the laughter of men, the roar of cars. The light of the elevator went up and down, it was a gigantic toy. On the other side was the sea, the moon and the stars, all lit up too. The music coming from it was sad and penetrated deeper. The sloops and canoes arrived quietly, the fish passed under the water. The noisier city, however, was much quieter. There was none of that at sea. The music of the sea was sad and spoke of death and lost love. In the city, everything was as clear and mysterious as the light of the lamps. At sea everything was mysterious with the light of the stars." (Amado, 1936, p. 54)

City and pier - land and sea - appear in the text as spaces filled with human life. The references to the city are to progress, noise, bustle and a certain fascination with the technology presented at the time. All the more so considering that the gaze of those looking at it was that of the seafarer. However, even though the city fills an imaginary that reveals awe, it is on the quays and in its continuity with the sea that the depth of Guma's existential relationship is born. The other side of the city was also represented in a reference to the sea: deeper, sadder, more unpredictable. The great preference for the sea as his point of admiration in the writing of *Dead Sea* comes from the revelation that the sea is the place where "music penetrates deepest", into the soul. Added to this is the realization that the city was noisy, but quieter. The sea holds the mystery of life for the sailor. It's where love lives.

The destiny of not only Guma, but all the real sailors *at the Dead Sea, is to* throw their lives into the sea. The sea is their road, their home, their dwelling place. The sailors, who are the Amadian men who live in search of adventure, can't find it on the land and in the safety of the city. Love, luck, the meaning of life - even if it comes through the arrival of death - are all associated with the sea. The land through the city would be his great misfortune and the escape from his true identity. The city embraces him in comfort and not in the temptation to seek what he lacks: the meaning of his existence.

As we move further into the interior of Bahia, the defined territory of the book, even with borders described in a diffuse or incomplete way, these dualities of land vs. sea begin to appear more explicitly as contrasts. Apart from the Hsical distance, the further inland the pages of the book describe, the more distant life on the edge of the pier becomes. The contrast is between the sertao and the sea. Another recurring theme in Brazilian literature .[4]

Among these works that portray the same sertao x mar duality we can mention: Os sertoes (1902) Cunha, Euclides; Romance da Pedra do Reino and o principe do sangue do vai e volta (1971) Suassuna, Ariano; among others.

An interesting observation about the duality of the sertao is that in this passage of the book the relationship of exaltation is reversed. This time, the memory of the sertao does not belong to Guma, the sailor. It falls to Livia, his wife, who fears the sea, and because of this, builds a relationship of fear and conflict with the sea. In Livia's imagination of the sertao, this space represents paradise and a desired place. Once again, a portrait is drawn in comparison to the sea. Livia, who fears for Guma's life at sea, idealizes a dwelling far away from him, in the sertao. Let's look at the following passage:

"She would only desire it, only love it completely if she could run far away from the sea that night. To go to the harsh lands of the sertao, to escape the fascination of the waves. The men there, the women there, live thinking about the sea. They don't know that the sea is a brutal lord who kills men. A song from the sertao says that the wife of Lampiao, who is the lord of all that, cried because she couldn't wear a dress made of steam. The steamer is from the sea and no one is in charge of the sea, not even a brave cangaceiro like Lampiao. The sea is the master of lives, the sea is fearsome and mysterious. Everything that lives in the sea is surrounded by mystery." (Amado, 1936, pages 154/155)

In Livia's mind, the sea is something to be feared and so life on the docks is not the most desirable. Livia dreams with all her heart that life would be much better if she went to live with Guma in the sertao. Then she wouldn't have to fear his death. For her, there is something worse on the quayside than the misery of the factories and the hard life of the sailors or in the fields: there is the certainty that death will come over the sea, on an unexpected night, completely out of the blue. Maria Clara, Livia's friend and a sailor's wife born on the docks, has no anguish in her heart because she knows that it has to be this way, that it has always been this way. She was born at sea, all her people are in the ocean. Only Master Manuel, her husband, still crosses the water. But Livia came from the land, she wasn't born in the sea, no one from her family stayed in the waters, no one went with Lemanja to the endless lands. Hence the difficulty in understanding the timeless dynamics of the pier and Guma's life at sea.

The converging point of this land vs. sea duality also emerges in the text. Although divergent in their rhythms and in the desires of their local inhabitants, the social and economic dynamics of these two habitats - the city and the quayside - are integrated.

Take a look at the following passage:

"The morning is beautiful, full of sunshine. October is the most beautiful month on this quayside. The sun isn't hot yet, the mornings are clear and fresh, they're mornings without mystery. From the nearby barges comes the smell of ripe fruit arriving for the market. Seu Babau buys pineapples to make delicious cachaga for the customers of the 'Farol das estrelas'. A black woman passes by with tins of porridge. Another sells mungunza to a group. Old Francisco takes two glasses of puba porridge. A sloop leaves loaded. Boats go fishing, the fishermen naked from the waist up. The market is bustling, men are coming down the elevator that connects the two cities, the upper and the lower." (Amado, 1936, page 165, 6p)

City and pier in Amado's work are complementary. The pier, located in Salvador's Lower Town, supplies the city, corresponding to the territorialities of the Upper Town in the same location. In other words, land and sea are intertwined in their economic functions, such as commercial relations; as well as in social relations, even though these further confirm inequalities in living conditions or relations of class and exploitation. Along the quay pass men selling fish, their rolled-up pants representing the sweat of their labor. It is the land of the fishermen. They are the link between the sailors and the city's population.

6 - The sea and the phenomena of nature

The sea is the sailors' road. It is wide and constant in the daily lives of Jorge Amado's men. Since sailors live by the sea, this road is one of the fundamental elements for understanding the paths that define these men's lives. The sea represents the path of their work, their home, their life. The destiny of the men in/of the sea is ancestral, hereditary. It passes down through the generations. From this perspective, the sea is a character and not merely a decorative backdrop to the plot. Like the sea, other natural phenomena are also presented in the text as "beings" capable of influencing or even defining the social dynamics of these seafarers.

In Amado, nature is alive in such a way that it gains the force of an autonomous and independent existence. It is capable of transmuting people's spatial and conjunctural dynamics

presented in that society in the novel. Look at the following passage about one of these possible phenomena of nature, the wind represented in the landscape:

> "The wind is the most fearsome of the dock's dominators. It roughens the waters, it likes to play with the sloops, to make them turn in the sea, dethroning the wrists of those at the helm. That night was his. He started by turning off the lanterns, leaving the sea without its lights. Only the lighthouse blinked in the background, showing the way. But the wind took them the wrong way, diverted them from their course, brought them out into the open sea where the waves were too strong for a sloop." (Amado, 1936, page 210, 3p)

The magnitude of the wind and its ability to transform is clear in this passage. The wind has a dominant force over the people and the sloops on the quay. The author mentions that the wind likes to play with sloops, that it dethrones the wrists of sailors at the helm. The wind is capable of blowing sailors off course. The wind is yet another character who defines the destinies of Amadian men.

The sea, a living and essential character in the geographical environment of this plot in *Dead Sea,* varies its strength and behavior greatly influenced by the other elements of nature associated with it. The strength of rain and thunderstorms is another phenomenon to consider. On stormy days and nights, the sailors themselves, as well as their wives, fear for their lives by going into the sea. The storm is the great adventurer when it reaches the ocean. It overturns sloops, dismantles ships, camouflages dangers within the sea. Two passages make this point clear:

> "It was July, the month of the south wind, of storms. In June, the lemanja is unleashed by the south wind, which

is a fearsome wind. It's very dangerous to cross the bar at that time and the storms are fearsome. It's the worst month for fishermen and sloop masters. Even the Bahians are in danger in June, even the big liners." (Amado, 1936, p. 138, 2p)

"The storm hit in the middle of the night. Usually that wind didn't bring storms, but when it did it was terrible. It hit in the middle of the night and caught many boats at sea. Guma was woken up by Old Francisco arriving from the 'Lighthouse of the Stars'. [They reached the mouth of Barra. Three sloops were afloat. The storm was trying to shipwreck those they had come to save." (Amado, 1936, p.197 4p)

The causes of the storms also represent the will of Germany. Another significant point in the argument for the thesis is when Amado says that "even the Bahians are in danger" in the first fragment quoted above. As well as associating the fact that a Bahian is a man of the sea, he proposes to say that even these most fearless of men are of limited courage in the face of the grandeur of the storms of the will of Lemanja.

In the second passage, Guma, the hero character, is summoned to try to mitigate the damage caused by the storm. However, even though he reaches the mouth of Barra, in the Bay of All Saints, the damage has already been done: destroyed sloops become part of the landscape of this scene at sea.

From the same natural phenomena, but in different situations, the lives of the people on the quay can actually be graced by lemanja and nature itself. In certain months of the year, with mild winds and few storms, and the coming of the sun, its presence is actually graced. Take a look at some of the other passages contemplating this new bias:

"They pulled smoke from the pipes. People came in and out of the Mercado Modelo. The sun shone on the small stones of the calgamento. A woman was spreading a towel in the window of a house. Sailors climbed onto the back of a ship and washed it. The wind began to rush, shaking the flying sand." (Amado, 1936, page 66, 2p)

With the presence of the sun, life on the quayside brightens up and much of people's daily lives become more fulfilling. Another point:

"The moon illuminates its route, o sea and a wide and good road. And to the northeast it blows, the fearsome northeast of storms. But now it's blowing like a friend to help you cross that stretch of river more quickly. The northeast brings the songs of the riverside, the songs of the washerwomen, the songs of the fishermen." (Amado, 1936, page 131 2p)

The moon illuminates, the sea is a road, the northeast blows like a friend. This friendly wind also brings social life, the songs of sailors and laundresses. The identity of the people on the edge of the docks of Bahia is being drawn with the elements of nature as the builders of their geographical existence. And yet another example of this bias:

"The night was hot on land. But in the sea there was a cool breeze that gave the bodies a dengue. In the sky of stars, there was a huge yellow moon. The sea was calm and only the songs that came from everywhere cut through the silence.[...] Leaving the sea, the sloops, your port. That's something that hurts a sailor, especially when the night is this beautiful, full of stars and with such a beautiful moon." (Amado, 1936, p. 233/4)

And finally, in this last passage, Amado makes clear this intertwined relationship between man's being and his existence associated with the sea in the case of the men on the quays. He clearly states that for a sailor, leaving the sea on a starry night with a beautiful moon is like leaving behind a bit of himself. And that's why it hurts, leaving the sea and its harbor with its songs for the sailors is like ripping a bit of oneself out of one's skin.

7 - The Dead Sea and transversal conflicts:

Jorge Amado was a writer whose extensive work was translated and read in countless countries. He gave more people a vivid and motivating image of the Brazilian people. Not only does Jorge romanticize the memories, sufferings and joys of our people, but he does so with some basic concerns: the search for beauty, freedom and justice. According to Darci Ribeiro, "he is our best expression of what is called engaged literature, of the highest quality, that is to say, lucid and socially responsible literature". (1997, p. 28).

Jorge Amado saw Salvador from the following perspective: a city divided between the cynical rich and the honest poor, over whom the red clouds of revolution hovered. And the context of his work, thinking here more specifically of *Dead Sea,* represents a writer whose value, however, goes beyond the pages he wrote, perhaps weighing more heavily on the social level of the literary universe. Amado himself said in an interview: "I am an old connoisseur of the history of Bahia, nothing more than that." (1997).

Considering these aspects about the writer Jorge Amado, the secondary conflicts present in *Dead Sea* are also conditional elements, albeit without the protagonism already revealed here, for us to understand the being of the man presented there. Considering the legacy that the writer left behind, we can point out in this book the teachings that he defends such as: having sex with black people is good; that for him it is through socialism that we will achieve less social disparity; the experience of religious syncretism; the complexity of racial relations and the importance of analyzing the ecological framework and the matrices of social domination implanted during the colonial period in Brazil and especially in Salvador, the first capital of Brazil.

Syncretism is characteristic of Brazil and one of the most recurrent factors in Amado's work. This mixture exists in such a way that we can't help but think about it both to understand our similarities and to affirm our differences. For Amado, the most striking element of Brazilian identity is "undoubtedly the mestigagem, the mixture. We are not this or that, we are everything: white, black,

Indian. This is what makes us unique and gives us real importance." (1997, p. 55).

In the work analyzed in this chapter, some of these elements of the identity of the Amadian work appear repeatedly throughout the narrative. However, some of them reflect secondary conflicts in the text, as well as criticisms of the status quo in force at the time of writing. Among them we can list: mestipagem and recurrent racial segregation. The strong presence of Arab migration. The denouncement of the social contrasts between the docks and the city; the miserable conditions of the exploitation of human labor. Let's take a look at some of these points.

One of the characters is Professor Dulce. She is responsible for bringing literacy into the lives of the people on the quays. Almost all the sailors were children born on the quayside and taught literacy by Professor Dulce. The teacher plays a prophetic role and clashes with the reality of the docks:

"And when the teacher saw them, she didn't recognize them anymore, they were men with bare chests and burnt faces. They still passed her by, shy, with their heads down, and yet they loved her because she was good and tired of everything she saw on the edge of the pier...[...] But she saw such sad things by the ships, in the rough houses of the fishermen, on the prows of the sloops, she saw the misery so close that she didn't have the courage and lost her joy, she no longer looked at the sea with the enchantment of the first days, she no longer waited for a fiancé, she no longer had rhymes for her verses. " (Amado, 1936, p. 50)

Even though life on the docks and at sea are viewed by the writer with a certain romantic bias, he also provides us with a moment of more realistic reflection. And this part of the story, which gives a clearer picture of the daily life of seafarers, was played by the character of Professor Dulce. When she arrives at the pier and sees the joy and attachment that the sailors have for all that life surrounded by the sea, the first feeling is admiration. The beauty of the sea, the adventures of the men of the sea, all this is very attractive compared to the hurried daily life of the city. However, the closer one gets to this everyday life, the more one participates repetitively in this day-to-day life, the dazzled emotion begins to crumble. Dona Dulce, who once only saw the beauty of life at sea, now sees the injustices of a life lived in poverty and misery. And that this is the most recurrent life on the edge of the quay.

To talk about Bahians and Bahia without mentioning mestigagem is to omit one of the main elements of the identity of this group and this place. Jorge Amado, the reliable declared teller of stories that reveal Bahia, even though he wrote many books on different themes, always made use of certain points that became obligatory in his narrative. Among them, the ethnic diversity that comes from mestigagem. It is common knowledge that black Africans are the founders of Brazil and its inhabitants. However, even if in smaller numbers, or coming through migration, other ethnic groups have arrived in the country and left the mark of their identities here, brought from their respective homelands. In *Dead Sea,* Jorge Amado also talks about the Arab colony that docked in the ports of Bahia de Todos os Santos. Toufick was the name given to this Arab character created by Amado. Take a look at a passage about him in the book:

"He had arrived in the third class of a lugger that had touched twenty ports. He'd arrived from the *other side of the world, carrying almost nothing in the leather wallet he'd clutched* to his chest as he began to climb the slopes of the Mountain. He had arrived on a stormy night, the night Jacques' sloop had capsized at the mouth of Barra. That night, on board third class, looking at the strange city in front of her, she had cried. He had come from Arabia, from a village among the deserts, overcoming seas of sand, to make a living on the other side of the earth. Others had come before him, some had returned and had olive groves, beautiful houses, they were rich. He had come for that too. He had come out from among the mountains, crossed stretches of sand on camels' backs, boarded the third class of a ship, lived at sea for many days." (page 218, 1p)

Toufick had come from Arabia. He faced "seas of sand" according to the writer. This is another interesting way of contrasting the world from which the Arab came and the one that draws him. However, the natural grandeur of this world is represented by the same shape as the world beyond the Atlantic in the waters of Bahia: the sea. Another interesting point highlighted by Amado is the strangeness of the city on Toufick's arrival. From a geographical point of view, it wasn't his "place" yet. He didn't feel like he belonged there yet. Still on the subject of the passage quoted, the sea takes on a new perspective: the first home of an immigrant.

The reference to the character Toufick - the Arab - in the book analyzed here, is perhaps one of the most geographical parts of Amado's definition of man. When he arrives, Toufick is a stranger to the place, to the environment. As his sense of place grows stronger, even though he is of a different nationality, even though he has not yet mastered communication, his relationships of belonging begin to build up. Let's look at the following passages on this point:

"He didn't even know the language yet and was already selling umbrellas, cheap silk, handbags and the maids and servants of Bahia. Little by little he became familiar with the city, the language and the customs. He lived in the Arab neighborhood of Ladeira do Pelourinho, where he left every morning with his peddler's bag." (Amado, 1936, p. 218)

"He knew his way around the quays with few people. The sloop masters were familiar to him, he knew all the names of the boats, even if he pronounced them in a picturesque way. [...] And with so many unknown and dangerous routes, Toufick, the Arab, was almost a true sloop master." (Amado, 1936, p. 219)

Toufick arrived in Salvador and moved to the Arab neighborhood of Ladeira do Pelourinho. Another sign of the observed migratory trend. He joined his "patricians" in order to better adapt to his new spatial reality. Little by little, familiarity was born, and even without mastering the language, Toufick, by repeating so many journeys, so many comings and goings to/from the sea, began to feel like a local on the quays of Bahia de Todos os Santos. A part of Toufick, from practicing "being" a Bahian/sailor, became one too. Associated with this reflection, let's ask ourselves what conditions, in the case of *Mar Morto*, the same identity for o Bahian, or o black or o Arab in the case of Toufick? What is the "what" that universalizes them? Amado answers us:

"And on a dark night he sang in his own language a sea song that he had heard from his sailors in the port where he was sailing [...] the songs of sailors, however different their language and music, always speak of love and death at sea. That's why all seafarers understand them, even when they're sung by an Arab from the mountains who heard them in a dirty port in Asia." (Amado, 1936, p. 219)

The sea is something that brings men together in a universal condition. The sailor/sailor from Salvador has little in common with the Arab Toufick in the eyes of others. Skin, hair, beliefs, religion, language... However, the sea unifies them both in terms of their working relationship as a condition of existence, as well as in terms of what becomes the essence of their daily lives and also defines them as being.

Another typical subject in Amadian works is racial themes. In *Dead Sea,* the oppression of whites over blacks in Africa, the continent visited by the sage who narrates the story, is briefly portrayed:

"In the parts of Africa where I live, my people, black life is worse than dog life. I was in the black lands that now belong to the French. There, black people are worthless, black people are only white people's slaves, they get whipped. And that's their land..." (Amado, 1936, p. 203)

The story told by the traveling sailor generated commotion and revolt. Anyone who heard it could identify with it, either because of their skin tone or because of the description of white exploitation. However, one element was the great differential of the blacks, even if they suffered, in Jorge Amado's work: here in Brazil, in Bahia, they already have their freedom. Even if it was restricted, even if it was conditioned by countless factors that were not their decision, the black Bahians of the docks were no longer someone's "things", as were their past generations. And for this very reason, their freedom was unquestionable. Africa, which was at the mercy of French colonization at the time of the book, is still suffering from the lack of black freedom.

The final point of conflict, which runs through much of the book as well as analyzing Guma's trajectory, is the economic crises that result in a very difficult life for the families on the quay. The capitalist economic crisis hits the docks:

"...things were bad for everyone on the quayside. They were so bad that the dockers were even talking about going on strike. Guma was always looking for work, making the trips as fast as he could to keep the customers. Several sloop masters sold their boats and took other jobs in the port: docks, long-distance ships, transporting suitcases and travelers' belongings." (Amado, 1936, p. 223)

The economic crisis that had shaken the pier affected several factors. Firstly, the possibility of the sailors losing their only possible asset: the sloop. With the crisis, there was little cargo to be transported. Schooners were also beginning to lose the battle to speedboat technology. Some began to look for alternative activities to find income, even outside the sea. Others even turned to illegal trading for lack of options. And, the crucial point of this part of the book: the class struggle of the workers appears in the text. The dockers, hit hard by the crisis, begin to organize strike movements. This was definitely a hallmark of Jorge Amado. Turning his literature into a commitment to social denunciations. For Vargas Llosa, "his public figure and his literary work were identified with the idea of the engaged writer, who uses his pen as a weapon to denounce social injustices, tyrannies and exploitation, and win supporters for socialism." (1997, page 38)

8 - The women at the Dead Sea

The great representation of the transformation of positioning and which marks some gender debates, also worked on by the writer throughout his work, is the case of Livia, Guma's wife, the female protagonist of the book.

Livia was a city girl. Her family weren't pier folk, but one day she met and fell in love with Guma. And when she married him and moved in with him, her destiny changed: she became one of Salvador's sailors' wives. Livia, a sailor's wife, feels the sisterhood that she now carries with all the other women on the quayside: having the same fate, of hoping that one day the news will arrive that the sea has taken her man away forever.

The great rite of passage described by Jorge Amado, in which Livia stops being a city girl and becomes a sailor's wife, is built on her marriage:

"Even sailors who travel the distant seas in huge liners come to get married in the church of Mont Serrat, which is their church, perched on a hill overlooking the sea. She got married there, to Guma, and ever since, on the nights of the quayside, on their sloop, in the rooms of the 'Farol das Estrelas', on the sand of the quayside, they have loved each other, their bodies confused over the sea and under the moon." (Amado, 1936, p. 26)

Guma and Livia's wedding was in keeping with seafaring customs. There was guitar, singing, dancing, cachaga, sweets and feijoada. And they sang the songs of the sea, from the one that says the night is for love to the one that says it's sweet to die at sea. Livia, however, began to understand better the dynamics of living on the edge of the quay and living the life of a sailor's wife. She began to fear for Guma's life and his journeys out to sea:

"His wedding march had been that cangao of doom. A cangao that summed up life on the pier. He drowned, any woman could tell when her husband left. Sad fate for her.[...] Her husband drowned every day in the green waves of the sea. One day his body would come instead, he would sail the Lands of Sim firn do Aioca." (Amado, 1936, p. 154)

Livia, once of the city, of the life of comfort and without the life-threatening adventures, couldn't get used to her husband's predicted fate. Guma, the bravest sailor on the quayside, would never see himself whole and alive without this relationship with the sea. And Livia knew it. The fate of the men of the sea is ancestral, hereditary. It passes down through the generations. Livia's apprehensions increase when she realizes that she is expecting Guma's child. Another son of the sea.

Livia, who couldn't accept losing her husband to the life of a sailor at sea, let alone her son, is facing a dilemma: convincing Guma that it's possible to have a safer and more comfortable life for the family in the city. On the other hand, she knows that for Guma to choose to give up his life on the quayside would be like not being able to live anymore. On Livia's thoughts:

"In the same way, Livia keeps waiting for him to come back. Surely she would like him to leave the sloop, go to the city and work in another profession. But she never mentioned it because she knows that men who live on the sea never go ashore to work in another profession. Even those who come to the sea, сото Dona Dulce, don't come back. The witchcraft of the lemanja is very strong." (Amado, 1936, p. 184)

Through Livia, Amado discusses the conflict between the dynamics of the sea (pier) and the land (city). For those of us who were born and raised on the edge of the sea, living away from it and chained to urban daily life seems like a great suffering due to the loss of freedom in the adventures of the boats. Livia, born on the asphalt slopes of Salvador, says that life at sea can be an attractive adventure. However, as long as it doesn't involve the risk of losing the lives of her husband and, in the future, her son. "Livia's days are sad with waiting and fear. The pier is beautiful, the sea comes crashing against the rocks, there is no more beautiful sky. There is music on all the sloops in the mouths of men. But for Livia, the days are sad and full of suffering." (Amado, 1936, p. 206)

Throughout her text and the feelings of Livia, who represents the women of the sea in the *Dead Sea,* the relationship with the sea is very conflicted. Sometimes hated, sometimes feared, but also sometimes loved. We have the following passage on these dualities:

"And she liked to go to the sloop on the nights when her son went for a walk to her aunt and uncle's house. She would lie beside Guma, listening to the songs of the dock, seeing the yellow moon, the countless stars, feeling the presence of Lemanja, who stretched out her hair in the water. I thought the sea was a sweet friend. And I felt sorry for Guma, who was leaving the dock, leaving his destiny. But he wouldn't sell the sloop; once in a while, when the sea was calm enough, they'd go for a walk on the water, look at the stars and the moon from the sea, listen to those sad songs from the pier. Then they would love once again on board the sloop. The waves would bathe their bodies, the love would be even better.[...] But the memory of their son growing up in the streets of the city, growing up for a better destiny, would console their hearts for the sacrifice they had made. But they would still miss him, they would miss him terribly, as one misses a loved one. Because no one can be born or live at sea without loving a lover or friend." (Amado, 1936, page 244)

The announcement of the great tragedy of Livia's life comes true: Guma ends up in the arms of the sea. Her great fear is realized. And once again, in describing the episode, the sea is once again the main protagonist in the scene of the characters' experiences:

"It was all right when Guma's body disappeared. Now the waters are calm and blue. Yesterday they were stormy and green. But to Livia's eyes the waters are still and leaden. It's as if the sea had died along with Guma." (Amado, 1936, p. 251)

The color of the sea represents the time of change for Guma and Livia. The variation of colors represents the state of mind and feeling in which the characters find themselves in the Amadian narrative. The sea, once again, is not just a component of the scene. Its dynamic colors, in this example, represent the different feelings that people are experiencing.

The big turning point in Livia's life is again associated with the life of seafarers on the quayside. As much as she had been brought up in the city and longed to return to a safe life with her son, without the oceanic adventures, she felt once and for all that she no longer belonged to this city context outside the pier (without a comma). Her spatial relationships, the way she lived and built her being, now necessarily involved life on the edge of the pier. The sea, which she had once feared, was now also her defining element of daily life. Livia watches the 'Paquete Voador' - Guma's boat - slowly bobbing on the waves of Salvador's sea. From then on, she felt that there was no point in running away from her destiny: her life now belonged to the dynamics of the pier and her relationship with the sea. About Livia and the 'Flying Paquette':

"She turns her head, looks at the 'Flying Paquette'. One of the best and fastest sloops on the quay. Few like it. How proud Guma was to say that! He loved his sloop, he had bought it for his son, he had died to keep it. Now she was going to sell it, she was going to give all that was left of Guma at sea to another man. It was like giving up her body, like letting herself be possessed by someone else. [...] Livia looked at the 'Flying Paquette' and felt a great love for it. Selling it was like selling her body. And they were Guma's things, she couldn't sell them." (Amado, 1936, p. 259/ 260)

Lfvia realized that through the 'Paquete Voador' she would remain in communion with Guma: it was the boat he loved, it was the boat destined for their son, it was the boat that took him to sea and kept him alive. Amado turns Lfvia into a heroine the moment she assumes once and for all her condition of life linked to the sea by the pier. Lfvia begins to see her existential condition as linked to the sea. It would be through the sea that she would remain with Guma, she would only find him, now in the arms of Lemanja, if she went to sea herself.

In another passage, once again Amado leads us to reflect on how the situation of the sea reflects the feelings of the characters. In this case, it was Lfvia's:

"Livia looks at the dead sea of leaden waters. A sea without waves, heavy, a sea of oil. Where are the ships, the sailors and the shipwrecks? Dead sea of floundering, where are the women who don't come to mourn their lost husbands? Where are the children who died on the night of the storm? Where is the sail of the sloop that the sea swallowed up? And the body of Guma who floated with long brunette hair in the water that was blue? In the feathery, heavy water of the oil-dead sea, the light of a candle runs like a haunting, looking for a drowned man. And the sea that has died, and the sea that is dead, that has turned to oil, has stood still, without a wave. A dead sea that doesn't reflect the stars in its heavy waters." (Amado, 1936, p. 262/263)

Livia's dead sea and the sea without Guma. This is the same sea that takes the lives of sailors and leaves the women and children on the quays helpless. For Livia, the sea is heavy, the waters are not green but leaden. The sea has lost the magic of reflecting the moonlight on nights for the love she shared with Guma. The outline of the sea and the way it is seen define exactly the feeling that suited the character. The women of the sea are those who love it just like Lemanja, or see it as dark, without color or life, like those who lose their men to the adventures they experience in it.

Jorge Amado concludes his *Dead Sea* in a great symbiosis: Livia is seen as a lemanja. Livia, in order to keep her and her son's bond with Guma alive, decides to take over the place of "marftima" with the 'Paquete Voador'. The boat that used to belong to Guma is now commanded by her, a woman, who takes on her new trade. Her goal is also to prepare her son to take his rightful place. This is Livia's new saga as a sailor:

"Livia suspended the sails with her womanly hands. Her hair is flying, she's standing up. Alcanga o 'Traveler without a Port', Master Manuel lets her go ahead, he will steer o 'Flying Paquete'. Seabirds circle the sloop, passing close to Livia's head. She stands up straight and thinks that on another voyage she will bring her son, his destiny and the sea." (Amado, 1936, p. 264)

Livia, representing the women of the docks, embraces her new destiny: she goes to meet the sea. In the city, her life would be restricted and dependent: her built devices were not for urban work, except domestic work. Livia had to work to support her son and her new life without Guma. The great solution was to take on the work role that Guma played in his family cycle. Running a boat business had always been considered a man's trade. And once again, the Amadian narrative brings a new reflection on gender and class: a woman can take on new positions. Taking this path, building a new life situation, Livia represents the strength of women who fight independently for their place in the sun. By seeing a strong Livia, a woman of fiber, Livia therefore assumes the vision of lemanja:

"Morning star. On the quayside, old Francisco swings his bridle. Once, when he did what no sloop master would do, he saw Lemanja, the mistress of the sea. And isn't she the one who's now standing on the 'Flying Paquette'? Isn't she? Yes, she is. It's Lemanja who's going there." (Amado, 1936, p. 264)

9 - The geography of man at the Dead Sea

Throughout the reading of *Dead Sea,* different geographies have been pointed out in the body of this chapter. O geographical knowledge, that is, "the doing of Geography" was born and developed from the social practices mentioned, constituting the scope of the processes of geographical foundation. In these notes on the book, Geography is permeated by the dominance of the subjective as an integral part of what can also be the objectified scientific knowledge of this science.

By reading Amadian literature as an empirical source for the scientific paradigm proposed here, there will be different perspectives on the meaning of the world. Geography or the geographical, as a foundation, is something that is established from the relationship between man and the environment. The sense of foundation is related to the Geography that we can see in Jorge Amado's work, which was analyzed in this chapter. However, from the plurality of definitions that these geographies can have, the Geography that is born from this empirical literary reading necessarily becomes plural in the same way.

The aim of this chapter is to develop a geographical reading of the world from an ontological perspective by looking at the *Dead Sea.* Geography can be recognized as a knowledge in which one of its purposes is to give us a way of conceiving things in the world. It measures "being in place"; "being there", "being in itself" in the world. We can reflect that each conception of the world involves a type of logic, and a type of experience/existence is expressed as a result of this in the content of reality. Any knowledge of a geographical nature that is constructed necessarily involves man's existential condition. In the book we analyzed during this period, man is defined by his relationship with the sea. This element is the foundation of Amadian man. It is from this perspective that the geographical foundation or geographies are born.

Reality would have a geographical foundation. Geography, or the geographical, as a foundation, is something that is established from the relationship between society and nature. This translates into a spatio-temporal order of the resulting elements. Jorge Amado's book translated into words the experiences perceived to this day - albeit re-signified and (re)located - in the streets and along the quayside of the city of Salvador. The geographic foundation is related to this moving sea that brings and carries people in their daily lives. The geographical reality lies in the different existences between the people of the pier and the sea and the daily lives of the more urban citizens of the city of Salvador. This geographical awareness is born out of the spatiotemporal determinations that make the existential conditions of these same characters possible.

Men organize their empirical lives according to social relations that exist in the environment, given in space-time, revealing geographies. It is therefore in this context that existence takes place, because the relationships announced there lead to the movement of being and non-being. And that's what existence is all about: being is always ceasing to be (Martins, 2007).

In this thesis, a look beyond systematically organized and institutionalized scientific knowledge concerns what we will call Geography. In order to make this proposal a reality, it will be the geographic foundation and the geographies that will be pertinent to the debate in this ontologically-driven look. The geographic foundation will be present both in the reality of man's existence and in the reality of man's existence, which is pointed out throughout Amado's text. Martins (2007) states:

> "Identifying the geographical basis of reality means clearly perceiving geographical phenomena. It's as if we were looking at the world and asking ourselves what is geographical about it."

In *Dead Sea,* Jorge Amado clearly does this exercise. He looks at the sea, at the people who live there on the quay and exist around it, and defines who they are. The apprehension of the reality of the world around us, via the senses and the description of the environment that is observed, bringing to light the geographical foundation, can be read as the birth of human consciousness. Being conscious is knowing how to (re)know this reality of the world that surrounds us. Jorge Amado was able to make this peculiar reading of Bahia de todos os Santos very lucidly. Consciousness becomes the result of all this reflective objectification of reality and our ability to grasp it.

Armando Correia da Silva (1978) states that Geography's true "laboratory" is our reality. Thinking about the empirics present in *Dead Sea,* and agreeing with Armando, we can read in Amado's sentences a writer of geographical developments.

In this ontological debate that is being built on the foundation of man's reality, we will incorporate the reflection on the "being of man". The elucidating explanation for this idea is to consider man's being as a historical process that sees its construction in a dialectical process. It is from a constant becoming that being and being are founded, remaking themselves and defining who man is. This becoming will be founded and will emerge throughout our reasoning from the category of work. It was the work of the *Dead Sea* men that defined them. Developing a new line of reasoning inspired by the ideas of Lukacs (1979), we can designate the man who works, that is, the animal who becomes man through work, as a being who gives answers. Indeed, "it is undeniable that all labor activity arises as a response to the growth that provokes it." (Lukacs, 1979, p. 5).

Following on from the previous reasoning, men are founded/recreated/self-created as men through work. Because of this, since work is the category/action that collectivizes/universalizes man, this being comes from sociability, man's being also has a social character.

The observed dynamics of the quayside dwellers configure not only the peculiarities of everyday life, but its ontological existential element. In this specific case, everyday life loses much of its authenticity outside the maritime setting. Hence the observation of the geographical landscape.

The realization of the immediate existence of the geographical foundation is given by the practice of geography. If beings are defined as men, humanity is the existence in/of the world. This existence in question also corresponds to man's being. Geography, in a scientific debate or not, will be understood as the category of existence through the fact that it deals with the existence of a man now defined by his being.

10 - The geographical environment at the Dead Sea

One of the geographical categories used throughout the text to extract the geographical foundation of Amadian literature in *Dead Sea* was the geographical environment. Through it, we make explicit with density the relationship between society and nature already mentioned. The geographical environment will be used as a theoretical resource for understanding the spatio-temporal nexus of the Amadian world, where the social relations of existence are constructed.

Geography is therefore born out of the relationship between society and nature. It is the property of humans and the property of the geographical environment. But it is above all the property of a whole, in which this relationship is the extension/projection of man and society insofar as they are the creation and representation of the geographical foundation of existence.

In order to reinforce the way in which the geographical environment has already been theorized, considering that it is the result of the relationship between society and nature, we would like to reinforce its definition based on the intersection between the natural environment and the cultural environment. It changes over time as natural and human history develop. And by analyzing its dynamics, as the elements of the geographical environment are related to each other, the natural or cultural links that unite them define a space of relationships.

The geographical environment, as we have already discussed, can be translated as the order imprinted on nature through man's sociability in which he constructs his geography and dialectically reconstructs himself as well. Let's take a look at another literary passage from Amadiano that translates this definition well:

"Now all he has to do is wait for an adventure that will take him away, and that will also take away the 'Valiant' that he would like not to pass on to someone else. Because it's impossible for him to run away from the dock and go to other lands. Only those who live at sea know how impossible it is to leave it. Even for those who can't look into the face of a friend or gaze at the bright moon in the sky." (Amado, 1936, page 185, 1p)

The dilemma of Guma, a sailor and our main character once again, was whether or not to stay in the life of the dock. However, he understands that life at sea also defines him. Although permeated by a sense of adventure, that environment was built from him and dialectically founded him too.

When analyzing the properties described about the understanding of the geographical environment, a certain degree of caution in interpretation is necessary. The hasty idea that the constitution of the geographical environment is a uniquely harmonious process in a state of total equilibrium would be a misinterpretation of this category. The contrasts embedded in the spatialities of the geographical environment are not only present and make up the reality that forms it, but are also necessary in the face of the otherness that exists to identify the world.

Man now emerges within the man/nature relationship of the geographical environment. Survival and existence require the relationship between man and the environment, between society and nature, in order for choice to be made, and for this to be defined in work. This, in turn, has already been pointed out in this text as an essential element in defining our theoretical arguments.

11- Amadian geographies in the Dead Sea

This category of geography will be the one most used throughout the thesis and will be the one that sews the links between the chapters of Jorge Amado's books to be analyzed.

What is this geographical foundation, the object of study of Geography, the ontological reading of the world called geography? Initially, geography will be present in the reality constructed through the relationship between man and the geographical environment. Furthermore, from now on we can also designate geography from the point of view of geographicity, since it becomes the existential foundation that gives rise to the adjectives that constitute being. The ultimate expression of man's being is concrete through the materiality of geography.

Martins (2007) attaches great importance to geography. He even states throughout his text how at various times Geography is understood as a synonym for geography. This, in turn, becomes tangible and its understanding is achieved through its description as a possibility.

The sense of the geographical can be seen as the reading of geography. An important consideration in line with this reasoning has been made about the scope of the sense of localization. Let's give an example.

Livia had a conflicted relationship with the sea. She loved it and feared it at the same time. Her reflections focus on this sense of geography. For her, these same questions would not have been asked in another place, in another spatial logic, in another part of the world.

In Dardel's (2011) reasoning, the subjective content of the analysis of reality and the scope of geography is much more emphasized. Geographicity is thus "love of one's native soil or search for new environments, a concrete relationship linking man to the Earth, a geographicity of man сото the mode of his existence and his destiny." (Dardel, 2011, page 1)

In this more phenomenological reading of geography, another important consideration is that land and man are not merely a complementary and defining pair of reality. Here, in an objectivation-subjectivation relationship, they will be read as an inseparable pair. One is founded and recreated in the other. They are conditioning existential elements. They cannot be thought of geographically separately.

Dardel inaugurated a new ontological perspective for geography. Man, in this defining nexus of geography, has a whole meaning defined in a very particular way. And, reading this man in Amado's work, in the writer's *Dead Sea*, even the very title of the book carries a much more latent meaning. Let's read a very Dardelian passage:

"Some time later, the storm calmed down. The moon appeared and Lemanja spread her hair over the place where Guma had disappeared. And she took him to the mysterious voyages of the mysterious lands of Aioca, where the brave go, the bravest of the dock.[...]It was right there that Guma's body disappeared. Now the waters are calm and blue. Yesterday they were stormy and green. But to Livia's eyes the waters are still and leaden. It's as if the sea had died along with Guma." (Amado, 1936, p. 251)

Guma is a man born and defined by the sea. And in his heroic and tragic ending, he follows his destiny: he disappears into the waves of the Lemanja Sea. The element of the sea, in agreement with Dardel, defines the existential dimensions of Guma's life and death.

In a third conception of geography, its foundations are defined in a spatial logic of reading geography. Moreira (2004) uses the expression "the being-of-man-in-the-world" to refer to geographicity, something that is concretized throughout the book about the life of Guma and all the secondary characters, culminating in Livia's ending.

Moreira (2004) also reads geography from the point of view of man's being. According to his reading, geography is in itself the synthesis of the relationship between the concrete essence and existence of being. He designates:

"Geographicality is the spatial condition of man's existence in any society. The equivalent of what in philosophy Heidegger calls the worldliness of man. Or, in another context, Hegel refers to man's being-in-the-world." (page 32)

Closing this chapter with this passage, throughout these pages the search and the fragments chosen have been an attempt to extract the geography of man in his logic of "being-in-the-world". In turn, this logic is so well practiced in Jorge Amado's narrative. *Dead Sea* is this book that so well intertwines the living materiality of everyday social life with its setting as an element that shapes and defines it. Guma is this protagonist whose being in experience and existence is tied to the dynamics of the sea. And Guma is not just a character with an individual connotation. All of his conflicts discussed here - whether or not they involve secondary characters - could perfectly fit into the life of a sailor living today in the city of Salvador. Expanding the scale - leaving the micro of the individual and broadening its scope - the sea is an element that is the content of what we could say about Salvador.

Chapter 2 - The Geography of the Tent of Miracles:

There is a genuine originality that makes us different from others. This originality, this national face, the ambient reality, makes the eternal exist without the temporal. And there is no work by Jorge Amado in which this existence has been better demarcated than in Tent of Miracles. In the case of Bahia, what is this fundamental mark of originality? It's the mixture. Bloods, races, religions, customs, blacks and whites, Indians and Mamelukes, rich and poor, mulattoes with mulattoes, mestigos with mestigas, and this skin color and this democratic awareness of being in that place emerged.

Jorge Amado's originality lies in the fact that for the first time people could see themselves in Brazilian literature with their own personality, in all their spontaneity as creators of culture. And something that will be very important to us at this point in the thesis is the realization that Amado's literature is not gratuitous, art for art's sake, but rather engaged, polemical, participatory literature. The author set out to talk about who we are.

Published in October 1969, *Tenda dos Milagres* was written by Jorge Amado between March and July of the same year. The author's favorite book, it takes up themes that had already been announced in previous works. Miscegenation and racial conflicts are relevant throughout the narrative. The novel offers us a range of reading possibilities, among which we highlight the dialogue established between the popular culture practiced in Salvador and which defines this place and its people, and the scientific discourse developed.

The period presented as the space-time of the novel would comprise the end of the 19th century and the beginning of the 20th, in Salvador, when in Brazil, in order to modernize, the so-called hygienist theories were applied as the basis for the creation of a national project: race differences and their evolutionism became the key point for the creation of the national ideal type. Different intellectuals developed theories about this context which, among other things, fostered the ideology developed in national teaching centers such as the Bahia Medical School .[5]

In the city of Salvador presented in the novel, we see the strong changes undertaken to mobilize the idea of progress through the installation of industries and urban development. And the generation of progress was a theme frequently addressed in other works by Jorge Amado such as: O Pais do Carnaval (1931), Capitaes de Areia (1937), Jubiaba (1934), among others.

Jorge Amado's novel *Tent of Miracles* is an instrument of vindication and defense of the cultural wealth of Brazil's blacks and mulattos, a manifesto against racism and oppression, published in 1969, a period when the military dictatorship was becoming even more violent and destructive. Faced with this oppressive context, the novel is a political act of reaffirming the power of the people.

---

[5] **The theories presented and defended by: Silvio Romero (1908) Oliveira Viana (1934) and Nina Rodrigues (1932), among others.**

1- The work: Tent of Miracles

The opening of the book, in its very first pages, sets out to announce the justification and explanation for the novel's name: the Tent of Miracles. A setting that once again in Amadian novels helps us to define and understand the complexity of the characters. A setting which, because of the existential role it plays in the plot of the book and in understanding who the people presented there are, can repeatedly be considered one of its main characters.

The plot of the story takes place in the Tent of Miracles, which is so important to the narrative that it gives the book its name. The Tenda dos Milagres, located in Pelourinho, was not just a place where almost all the characters in the story lived and met. It was the place where the friends worked, the stage for capoeira de Angola, the writers of Cordel Literature, the setting for Candomble religious rituals, the mold of their lives. The Tent of Miracles was the free university of popular culture in Salvador, Bahia.

First, Amado gives us the spatial context of the Tent by introducing us to its location:

"In the vast territory of Pelourinho, men and women teach and study. A vast and varied university, it extends and branches out into Tabuao, Portas do Carmo and Santo Antonio beyond Carmo, Baixa dos Sapateiros, the markets, Maciel, Lapinha, Largo da Se, Tororo, Barroquinha, Sete Portas and Rio Vermelho, everywhere where men and women work metals and wood, use herbs and roots, mix rhythms, steps and blood; in the mixture they have created a color and a sound, a new, original image." (Amado, 1969, page 15)

0 Pelourinho is the territory of this so-called "popular university of Bahia": which included, in addition to the Tent of Miracles, various other specialized activities carried out by this territory based on popular territorialities. The concept of "popular university" is assertively defined by Amado. The author defined this place as an environment for maintaining all the popular practices that kept alive the Bahian identity established in Pelourinho. In addition, all the professionals who were in some way involved with Bahia's mestizo culture. Let's look at the following passages from Amado's book (1969):

"The teachers are in every house, every tent, every workshop. In the same building as Budiao's school, in an internal courtyard, the Afoxe dos Filhos da Bahia rehearsed and prepared for the parade, and Terno da Sereia has its headquarters there, under the command of mogo Valdeloir, a great performer in pastoral and carnival revelry." (page 16)

"[...] The large patio was also home to the samba roda, on Saturdays and Sundays, where the black Ajaiu was established, the only and absolute samba roda, its main rhythmist, its greatest choreographer." (page 17)

"There are many miracle scribblers, bringing them in oil, water and glue paints, colored pencils. Anyone who has made a promise to Our Lord of Bonfim, to Our Lady of the Lamps, to any other saint, and has been answered, deserved grace and benefit, comes to the tents of the miracle scribblers to commission a painting from them to be hung in the church, in grateful payment." (page 17, 2p)

"Troubadours, viola players, repentistas, authors of small brochures, composed and printed in the printing house of master Lidio Corro and in other deprived workshops, sell romance and poetry in the free territory for fifty reis and a toast" (page 17, 3p).

"In the Tent of Miracles, Ladeira do Tabuao, 60, is the rectory of this popular university. There is master Lidio Corro scratching miracles, moving magical shadows, digging rough engravings into the wood; there is Pedro

Arcanjo, the rector, who knows? Bent over old worn-out types and a capricious printer, in an archaic and impoverished workshop, they compose and print a book about living in Bahia." (page 20)

Teachers, writers, cordelists, sculptors, capoeiristas, musicians, singers, representatives of the terreiros, fathers and mothers of saints, violeiros, repentistas, popular actors, circus performers, rhythm players, percussionists, and many other references walked and frequented the slopes of Pelourinho on their way to this popular university with a rectory in the Tenda dos Milagres. And so, this geographical environment is being designed as men are founding and dialectically reconstructing themselves in the experience of this space.

Pedro Arcanjo is the main character in *Tenda dos Milagres, a* poor black intellectual who personifies Brazil's ethnic and cultural formation as he fights for the Bahian people to preserve their popular origins. Arcanjo defends miscegenation, fights racism and values the mestizo character of his culture. About the facets of Pedro Arcanjo:

"Archangel went out to see the world. Wherever he went, he learned. He didn't choose a job - grunt, bar tender, bricklayer's helper, letter-writer sending news and nostalgic greetings to the far reaches of Portugal. He wandered from mecca to mecca, always on the lookout for books and mistresses. Why was he so attracted to women? Perhaps it was because of his innate delicacy and easy words. He didn't just impose himself on women; he was still so young and everyone was already listening quietly and attentively." (Amado, 1969, p.115/116)

In this first passage, we see the side of a young Archangel, adventurous and not very fond of formal responsibilities. And an interesting reading of this period of his life was the difficulty of settling down in space. Archangel was of the world, always awakening to new discoveries. The Archangel of youth was the man who imposed himself by making everyone listen to him attentively. Archangel made the world his school, learning and remaking himself through his experiences. Arcanjo universalized himself even in the midst of his extremely local identity. The coming and going, the displacements, spoke of who Pedro Arcanjo really was.

With writing and literature, Arcanjo's way of being changes. Arcanjo's maturity fixates him more on his land, on his local existence. The more Arcanjo became involved with the theme of who this mixed-race person who lived in Salvador was, in other words, who his people were around him, who acted as a mirror to him and at the same time as someone to be segregated from the non-mestizos, the more he became rooted in his origin: Bahia, the city of Salvador, the Tabuao slope in Pelourinho, the address of the Tent of Miracles. His humanization increasingly took the shape of his place:

"Until then, his life had been all about the revelry of ternos, samba circles, afoxes and capoeira, the obligations of candomble, the pleasure of conversation, listening and telling things, and above all the laborious work of bedding women from one place to another in free diligence. Now his curiosity was no longer vain and free, leading him to candombles, afoxes, ternos, blocos, capoeira schools, the houses of old uncles, in long conversations with older ladies. An almost imperceptible but qualitative change, Arcanjo had acquired a complete awareness of the world and of life." (Amado, 1969, p. 181)

Arcanjo became a hybrid figure. And this hybridity, represented through erudite and popular elements, in reconstructing the character's human meaning, becomes over time his most coherent ontological element in his life's trajectory. At the same time, throughout this personality of antagonistic traits, Amado brings what he wanted to place at the heart of the debate: how to combine and bring together the erudite and rational science with mysticism and popular knowledge? The figure of Archangel personifies this. Take a look at the following two passages about his career:

"During that long decade, Pedro Arcanjo read up on anthropology, ethnology and sociology from what he could find in Bahia and what he had brought from abroad, collecting his own and other people's pennies.[...] At first, he had to clench his teeth to continue reading up on confessed racists and, even worse, the ashamed ones. He clenched his fists: theses and statements sounded like insults, they were slaps, beatings with a whip.[...] Because time had passed and the accumulation of knowledge had given Arcanjo serenity and security - he could see the foolishness where before he had suffered insults and attacks." (Amado, 1969, p. 229, 2p)

"He didn't abandon the pleasure of books for the pleasure of life, the study of authors for the study of men. He found time for reading, research, joy, celebration and love, for all the sources of his knowledge. He was Pedro Arcanjo and Ojuoba at the same time. He didn't split himself in two, with an appointment for one or the other, the wise man and the man. He refused to climb the small ladder of success and reach a rung above the ground where he was born, the ground of the hillsides, the tents, the workshops, the terreiros, the people. He didn't want to climb, he wanted to walk forward and he did. He really was Archangel Ojuoba, one and the same." (Amado, 1969, p. 229/230)

The book portrays a young, mature Pedro Arcanjo at the end of his life and how his legacy came to represent his image in multiple ways. And its definitive version could draw a parallel with the very mix of ragas he sought to defend and understand: Arcanjo's multifaceted vision was what would best define him:

"Suddenly the slope began to liven up. From Largo da Se, from Baixa dos Sapateiros, from Carmo, men and women emerged, hurried and distressed. They weren't coming for the death of Pedro Arcanjo, the wise author of books on miscegenation, perhaps definitive, but for the death of Ojuoba, the eyes of Xango, a father of those people." (Amado, 1969, p. 47)

The book reveals this ideological struggle between popular culture and European science, which was adapted through the various clashes between Pedro Arcanjo, delegate Pedro Gordilho and Professor Nilo Argolo, a professor at the Faculty of Medicine who was averse to miscegenation. Or even, in the fight carried out by the state apparatus through the newspapers and the police, in an attempt to minimize, or even exterminate, the cultural practices developed in candomble terreiros, capoeira circles and afox groups during Carnival. One of the most emblematic events in the novel is the organization by Pedro Arcanjo and Lidio Corro - his great partner and friend - of a clandestine afoxe parade in which they took part, showing the power games that take place in Salvador and ratifying this ideological struggle. During the Carnival period, the afoxe bloco had been banned. This confirmed yet another practice of segregating popular culture, the identity of the people and the destructuring of a social class that was already economically and socially repressed. Even so, the parade went ahead in protest:

"The whole Carnival came out to salute the Afoxe dos Filhos da Bahia, to applaud the liberating Republic of Palmares.[...] The people applauded the insubmissive, brave challenge; where have you ever seen, Mr. Doctor Francisco Antonio de Castro Loureiro, acting police officer and white man with a black ass, where have you ever seen Carnival without Afoxe, the toy of the poor people, the poorest of the poor, their theater and their ballet, their representation? The misery, the lack of food and work, the illnesses, the bladder, the cursed fever, the malady, the dysentery killing the children, you still want, doctor, to impoverish them more and reduce them." (Amado, 1969, p. 90)

The result of the parade generated various reactions: there were cavalry, police, running, fights, shouting, laughter, joy and the strength of popular culture. There was the press of the conservative elites, which published a note about the event stating that "the authorities should ban these batuques and candombles, which, in large numbers, spread through the streets on these days, producing this enormous noise, without tone or sound, as if we were in Quinta das Beatas or Engenho Velho, as well as this masquerade dressed in a skirt and torso, chanting the abominable samba, because all this is incompatible with our state of civilization" (Amado, 1969, p. 93). 93).

Another important feature of our scientific quest is the realization that Jorge Amado's language is yet another existential ground. The language he uses is that of his people, and the creator of the characters is surrounded by them, identified with them, listening to their speech at every moment. This is the way he has created stories and people about the existentiality of Salvadoran Bahia. The way in which Jorge expresses his intimacy with the themes of his characters seems to us to be done naturally, with a great deal of identification. Jorge's characters in *Tenda dos Milagres* speak "Brazilian". And in constructing this way of talking about the people, he is also defining and recreating them. This "Brazilian" speech is portrayed in the characterization of everyday life in Tenda dos Milagres:

"[...] There, ideas are born, grow into projects and are realized in the streets, at parties, in the terreiros. Relevant issues are debated, such as the succession of mothers and fathers-in-law, foundation songs, the magical condition of leaves, ebos and spell formulas. There, ternos de reis, carnival afoxes, capoeira schools are founded, parties and commemorations are arranged and the necessary measures are taken to ensure the success of the washing of the Bonfim Church and the gift of Mae Dagua. The Tent of Miracles is a kind of Senate, bringing together the notables of poverty, a large and essential assembly. There, iyalorixas, babalaos, literati, santeiros, cantadores, passistas, mestres de capoeira, masters of art and crafts meet and talk, each with their own merit." (Amado, 1969, p. 117p)

The whole Bahian or Afro-Bahian existentiality, of a raga that is being formed, of a mestizo civilization, is an integral part of every chapter of this novel and of every moment of the presence of its characters. Antonio Olinto (1972), in a relevant text analyzing the work, states:

"Jorge Amado's message in Tenda is that we need to defend the mixed culture we created and which makes us different from the vast majority of people today. It is within this mixture, through it and by it, that we realize ourselves as a people and as a nation. It is within this mixture that we become capable of making a new contribution to the civilizational movement of man, the humanization of man."

The magic realism *of Tenda dos Milagres,* practically on every page of the book, is even part of what human being is constructed in that spatial cut-out that is the city of Salvador for Amado. In this magic realist novel, the planes blend together, or rather: the reader can hardly tell if what they are reading belongs to the immediate layer of real truth or if it is linked to the world that pure reason cannot explain. What is interesting and relevant in *Tent of Miracles*, and what enriches its narrative, is that a Cartesian person can enter that space without understanding its existential meaning, yet, even if they don't believe in magic, they know that it has a usefulness that goes beyond its own meaning.

Around the environment in which reality and unreality are confused, a whole bunch of characters are stirring, from yesterday and today, professors from the Faculty of Medicine in times of racial priorities, directors and editors of newspapers today, male and female students, researchers from other lands, young poets looking for space in the supplements, real and fictional people, orishas and gods, and in the middle of it all, the figure of Pedro Arcanjo stands out with absolute clarity, an ordinary and wise man from a city, a researcher of the people.

The story is set in 1968, and the arrival in Salvador of Nobel Prize winner James Levenson causes a stir in the local press. The American professor came in search of four books documenting the formation of the Bahian people, written by a certain Pedro Arcanjo. We then return to the beginning of the 20th century, the time when the exploits of the poor, brown, bohemian and womanizing Arcanjo took place. But his theories, which valued miscegenation, aroused the hatred of Professor Nilo Argolo, for whom the mestigos were "degenerates". James Levenson's words on Pedro Arcanjo:

"...I didn't come to Bahia to talk about Marcuse. I came here to get to know the city where a remarkable man lived and worked, a man of profound and generous ideas, a creator of humanism, your fellow citizen Pedro Arcanjo.[...]In one of his books, Arcanjo wrote: 'the beauty of women, of the simple women of the people, is an attribute of the mestizo city, of the love of the ragas, of clear manners without prejudice. (Amado, 1969, p. 30)

Levenson came to Bahia to experience empirically what had aroused his scientific curiosity when he read Archangel's writings. Systematic in his purpose, he turned down invitations from academies, institutions, gremios, cultural councils, professors - all of which he had plenty of in New York - and "he was fed up, but when would he get that Brazilian sun again? On the beaches he even played soccer and was photographed shooting at goals, although women were undoubtedly his favorite sport. He "became intimate with some of Brazil's finest, on the beach and in the nightclubs." (Amado, 1969, p. 32).

The arrival of the American academic Levenson serves to reinforce the first criticism that Amado will work on throughout the text: the professor wanted to learn more about Pedro Arcanjo, a simple man of the people, whose published work on the traditions and families of Bahia had inspired him and aroused his scientific curiosity. On the other hand, the mestizo Bahian population portrayed didn't even know about his work, and academics of the so-called white lineage, for their part, only began to recognize the value of Arcanjo's writings after the movement of the American foreigner. Observe:

"The work of Pedro Arcanjo, the four small volumes on Bahian popular life, published with great difficulty, in minimal editions, in the precarious manual workshop of his friend Lidio Corro, on Ladeira do Tabuao, this work whose merits excited the American savant, was as ignored and non-existent here as in the rest of the country." (Amado, 1969, p. 36)

Amado, through the writings of Levenson's commentaries on Arcanjo, also raises another interesting issue: the universalization of elements of Brazilian culture. Let's take a look at this passage:

"[...] Master Pedro's phrase in 'Apontamentos sobre a Mestigagem nas Familias Baianas': 'If Brazil has contributed anything worthwhile to the enrichment of universal culture, it has been with miscegenation - it marks our presence in the collection of humanism, and our greatest contribution to humanity'." (Amado, 1969, p. 141)

Brazilian miscegenation, as presented by Amado, is in fact a significant element in the identity of our culture.

Pedro Arcanjo and Ojuoba, the eyes of Xango, the father of the disinherited people of Bahia. Of the blacks, the mestigos, the poor. King of the terreiro, of the afoxes, chamego of the women, friend of his peers, confidant and advisor to those who seek him out. Pedro is the mentor of Tenda dos Milagres, a house of knowledge, a printing workshop, a show house, in short, a quasi-popular university located on the slopes of Tabuao .[6]

The definition of the man in Pedro Arcanjo definitely passes through the other banner raised by Amado: the defense of freedom to worship Afro-descendant religions in Salvador and in Brazil. When Pedro Arcanjo is portrayed by the author as the eyes of Xango, he emphasizes yet another defining element of the character, and throughout the narrative, the culture of the terreiro is highly portrayed. The terreiro is one of these settings that help us understand the essence of Archangel, the Ojuoba[7] . About its dynamics:

"Umbe oxire fun ipako to Ijenan, there was a party at Terreiro de Ijenan. It was a big party, for Ogun, and a world of people came to see Ogun dangar. Ogun Aiaka danced beautifully to cheer up the eyes of the people who were tired of suffering so much.[...] From the door, Ogun said this to the soldiers: whoever is of peace, come to the Terreiro and dangle at my party. For my friends, my heart is honey of flowers, but woe to my enemies: for them

[6] **Ladeira do Tabuao is a popular street in Salvador's Pelourinho neighborhood.**

[7] **Ojuoba, a word from the Yoruba language meaning Eyes of the King or Eyes of Xango, is an Oye (African honorific title given to those who become high priests and dignitaries of the cult of Xango in Africa).**
in the candombles of Brazil).

my heart is pogo of poison." (Amado, 1969, p. 268/269)

In relation to the theme of Afro-descendant religions, which have definitely had to fight for their space to this day, a political clash is portrayed in the book: there is a territorial dispute over political influence in the city. The terreiro was the functional center of both the defense of black culture and the struggle for the rights of mixed-race peoples. Thus, in events involving territorial disputes, the police - legitimized by whites - persecuted Salvador's candomble terreiros. The territorial clashes took place, but Arcanjo's struggle went further: he dedicated himself to going against the system by writing a book about the traditions of the peoples of Bahia. Arcanjo wrote and Lidio Corro disseminated the publication. This is how it ended up in the Columbia University Library and straight into Levenson's hands. About this polemic fact, observe the excerpts below:

"The holy war, the civilizing crusade, went on for many years. During the reign of Pedrito Gordo, dandy and delegate, a bachelor with his reading and theories, violence was daily, without appeal or protection. Doctor Pedrito had promised to put an end to the sorcery, the samba, the negralhada: 'I'm going to clean up the city of Bahia'." (Amado, 1969, p. 275/276)

"From 1920 to 1926, during the reign of the all-powerful auxiliary delegate, customs of black origin, without exception, from the food sellers to the orishas, were the object of continuous and growing violence. The delegate remained willing to put an end to popular traditions, with a club and a machete, with a bullet if necessary.[...] the samba de roda was exiled to the end of the world, lost alleys and hovels. The capoeira schools closed their doors, almost all of them." (Amado, 1969, p. 304)

The narrative interweaves the erudite and popular registers with extreme skill. Criticism of the repression of candomble and other manifestations of black culture takes shape in the narrative of *Tent of Miracles*. And the key point in all this debate about miscegenation, constructing a geographical environment that is permeated by this conflict, is Tenda dos Milagres itself. A place in Pelourinho, the much-talked-about historic district, where the types so characteristic of its characters still seem to be lurking in every alley, on every hillside in the center of Salvador. It's one of the most important points about this repression. In order to disintegrate the strength of the black culture represented in the candomblé terreiros, they were pressured to leave the center, moving to peripheral and socially segregated areas. See:

"Many babalorixas and iyalorixas took axe and saints far away, expelled from the center and the surrounding neighborhoods to the distant rogas, places of difficult access. Others took the orishas, instruments, costumes, itas, songs and dangas, baticum, rhythms and moved to Rio de Janeiro - this is how samba arrived in the then capital of the country, in caravans of fleeing Bahians." (Amado, page 303/304)

The struggle for the continued freedom of Afro-descendant cults was very intense throughout the book. And, once again, the figure of Arcanjo was very important in this outcome: with his publication of the book on Bahian miscegenation, coupled with political awareness for the free space of blacks and the religious rites of candomble, together they managed to undo the holy war. As a final result, "the candombles were able to reopen the doors of their terreiros, the afoxes returned to the streets, samba spread in the carnival, ranchos and ternos, bumba-meu-boi and pastoris were reorganized. The capoeiras in the berimbaus and in the songs." (Amado, 1969, p. 312)

Turning to the present day, there is nothing more realistic than this spatial dispute over racial issues involving freedom of expression. The book's fictional, physical confrontations unfold in a real-life dispute, where some of these events are still very much alive today. A friend of Arcanjo's says that "they never stopped fighting, compadre: in the street and in the Terreiro, in the book and in the newspaper, in ink and in stone, at the party and in the noise. The longest fight, the dullest fight. Will it ever end, my compadre?" (Amado, 1969, page 313).

Another important point in the formation of this Amadian man in *Tent of Miracles*, highlighted here in the figure of Arcanjo, is his personal relationships. In his youth, Arcanjo met Lfdio Corro, a "miracle scribbler", who became his partner in the fight against racial and religious prejudice. Tenda dos Milagres, in Pelourinho, was the place where the friends worked:

"In the half-light of the late afternoon, in the purple light of twilight, Master Corro, sincere and moved, admires the finished work: a beauty. Another masterpiece to come out of this workshop, the Tent of Miracles, where a modest but competent artist struggles and strives in his trade. And not only in this trade of scratching miracles, in this art of painting ex-votos: also in many others, just ask in the street who Lfdio Corro is and how much he invents and accomplishes [...] there are two of them. Lfdio Corro and Pedro Arcanjo, almost always together, and with them together nobody can: compadres, brothers, more than brothers, they are mabagas, they are ibejes, two exus loose in the city. If you want to know, go to the police and ask Dr. Francisco Antonio." (Amado, 1969, p. 104)

It's even more interesting to realize that intrinsic to their relationship and the very definition of who they are to each other is the geographical environment represented by the Tent of Miracles. That roof is what unites them, what generates their complicity. For both Lidio and Arcanjo, their geographies are directly linked to each other and to the territorialities of that tent.

Lidio and Arcanjo's relationship was one of sponsorship for some of the children born and raised in Pelourinho. Bringing together all the knowledge perpetuated in the popular university of Tenda dos Milagres, they were both committed to providing studies and the possibility of social advancement for some of them. At times they struggled with the privileged classes, and these struggles had very serious consequences: such as the end of the Tent and the imprisonment of Pedro Arcanjo with his publications. And in these clashes, their friendship, sharing the same ideology, further strengthened their resolve that skin color should not be an obstacle to opportunities to improve one's life. Watch the passage:

"What are we fighting for, compadre Lidio, my good comrade? Why are we here, two old men with no money in our pockets? Why was I arrested, why was the printing press destroyed? Why? Because we said that everyone should have the right to study, to get ahead. Do you remember, compadre, Professor Oswaldo Fontes, the article in the Gazette? The blacks, the mulattoes are invading the colleges, filling the vacancies, and we need to put the brakes on, put a stop to, ban this disgrace. It became a feature article and the pages of the newspaper were pasted on the walls of the Terreiro. Tadeu left here, he started his climb here, he went up and he's no longer from here, my good, he's from Corredor da Vitoria, from the Gomes family, and he's Doctor Tadeu Canhoto." (Amado, 1969, p. 344)

From his communist convictions, which permeated much of his career, Amado was always an advocate - or even a defender - of reducing inequalities between social classes. These conflicts in which this debate was exposed are a constant theme in his novels. In *Tenda dos Milagres,* the space of opportunity for mestigos, blacks and the poor is created. And one of his characters, Tadeu Canhoto, a boy brought up in that environment and who studied with the help of the two, experiences the change in his social condition.

2 - Tent of Miracles: the geographical imagination of miscegenation

Tenda dos Milagres draws a parallel between the period of a Brazil Republic in search of a national identity and the Military Dictatorship of 1969, when the book was written. In this context, the city of Salvador became part of a hybrid and changing reality, where miscegenation was a pertinent debate in both periods. By examining the raga debate and the concept of nagao throughout the narrative, Amado brings us closer to an understanding of the city of Salvador.

Sellman (2013), in his thesis on the city of Salvador analyzed through the prism of Jorge Amado's works[8] , states that the most influential definition of nagao in contemporary society is the "imagined community". This, in turn, is conceived by the author as an idea of sovereignty through camaraderie, in a context where both the restrictions of the official religion and the prevailing economic pressures demean the territory and make this understanding of community necessary. Religion, in this specific case, can alienate a wider cultural system. Sellman is emphatic in arguing that the horizon of comradeship between peoples is effective in building a nation. And, within the Amadian context to be discussed, there is no denying the latent power of camaraderie in *Tent of Miracles.*

For many, Brazil's miscegenation, in long theoretical arguments with a scientific approach, is a major explanation for the country's insufficient development. Within this context, Amado embraced a debate in defense of miscegenation in *Tenda dos Milagres*, located in the city with the largest black and mixed-race population in the country, and with a strong political stance. In this way, Amado facilitated the debate on miscegenation in the country.

Medical schools - such as the Bahia Medical School discussed in *Tenda dos Milagres* - were engaged in adapting theories to reveal the causes of the country's backwardness, very much based on the theme of racial differentiation and miscegenation. However, some intellectuals at the time of

[8] "The construction of a city: Salvador in the writings of Jorge Amado". Thesis defended at the University of Manchester for a PhD in Philosophy in 2013.

this discussion of Amado's book argued that continued miscegenation could paradoxically dissolve this issue: somehow leading to the whitening of Brazilians would contribute to the gradual improvement of Brazilian society .[9]

Trying to grasp the look of the geographical imagination, greatly reinforced by this idea of the "imagined community" presented, and coming across the urban space of Salvador, one will find these geographic foundations in the lines of *Tenda dos Milagres*. The dominant perception of Salvador, which is still current, as well as its concept of the city, is presented in this concept of community in a pertinent way.

In order to contribute to this debate on miscegenation, we cannot omit the importance of Gilberto Freyre's book *Casa Grande & Senzala (1933)*. This book was very relevant to the debate on Brazilian identity, from genetics to culture. *Casa Grande & Senzala* unraveled the rural patriarchal system that dominated most of Brazil's society, politics and economy at the time, and Freyre's next question, which continued the debate in *Sobrados e Mucambos (1936)*, was to examine how the elements of this rural system were transplanted to the growing urban sphere, generating reflections that are among the propositions we are analyzing in this chapter of the thesis. For Sellman (2013), Freyre's greatest contribution to Brazilian culture was his view of miscegenation in the country, seen as a "reason for national optimism".

Casa Grande & Senzala was a turning point in this debate on miscegenation, which took shape in Brazil in the historical context of the publication of *Tenda dos Milagres*. In his pages, Freyre (1933) stated:

> "It's not that in the Brazilian there are, as in the Anglo-American, two enemy halves: the white and the black; the former slave and the former slave. Not at all. We are two fraternizing halves that see each other enriched by different values and experiences; when we complete each other into a whole, it won't be by sacrificing one element to the other."

Amado stated more than once in interviews that he had read and admired Freyre's work. The Bahian writer met the sociologist at a Congress of "Afro-Brazilian Studies" that Freyre himself organized in Recife (1934) to debate the theme of miscegenation and the formation of Brazilian identity. There are many things in common in the works of Amado and Freyre. Both set out progressive ideas about society and were also influenced by the international historical context of their times, adhering to many ideas from this period. Both Amado and Freyre came from rural areas in the northeast of the country. They were persecuted and exiled for the ideas they defended. However, the similarities of thought stop there, as their respective involvements with the politics of the time were antagonistic. After all

What is the point of this debate about the proximity between Amado and Freyre? The important thing

[9] **Schwarcz, Lilia.** *"The Spectacle of the Ragas", 1993.*

to realize is that the background to the ideas discussed in *Tenda dos Milagres* is reminiscent of the racial theories of the early 20th century.

Amado repeatedly paid attention to the theme of miscegenation in his novels. He was able to describe places in Salvador where miscegenation is an integral part of the detailed landscape. And often, this way of describing Salvador from a racial perspective also served to denounce the great spatial segregations inherent in the city. This is how his geographical imaginary is characterized.

In relation to the characterization of the geographical environment pertinent to this theme, while Freyre places the senzala as the stage for the events of the black and even mestizo population, Amado illustrates the geographical environment in a very different way: the terreiro is the great dwelling place of the blacks in Amado's novels. As we explore the city of Salvador through Amado's eyes, places like Tenda dos Milagres represent the scene not only of experiences, but of an ontological perspective on black people and their miscegenation. Salvador for Jorge Amado is a city with a segregated space, but at the same time, a segregated area that lives on the threshold of struggles against social discrimination, building an oppressive system that is still observed today.

In *Tenda dos Milagres*, this question of miscegenation and the writer's search for an "ideal" model of place for a more egalitarian urban society are addressed throughout the text. Amado does not deny his preference for portraying the lower class community, because he believes that it is among them that the true idea of Brazilian identity lies. The lower class neighborhoods - such as Pelourinho - are always of great importance in the development of the narrative.

A very interesting point in the ontological construction of Pedro Arcanjo is that Amado managed to merge popular and erudite elements in him. Thus, the Pelourinho and the noble part of the Upper Town coexist in the same figure of Arcanjo. He is a mestigo, and his existential formation process is linked both to life experiences on the streets of Salvador and to academic experimentation. In addition, the structure of telling the story of *Tenda dos Milagres* through flashbacks allows the author to recall various important historical events both in the formation of the city of Salvador and in Brazil, where the Bahian city of Pedro Arcanjo is conceived and perceived. The author of *Tent of Miracles* states through Arcanjo's speech:

"If Brazil has contributed anything worthwhile to the enrichment of universal culture, it has been with miscegenation - it marks our presence in the collection of humanism, and our greatest contribution to humanity." (Amado, 1969, p. 141)

Blacks, Indians and whites are the ethnic groups that make up our racial ancestry. Repeated assumptions are made about our miscegenation in Brazil, and almost unanimously, it is said that the idea of a Brazilian purely from one of these races is a fallacy when investigating the genealogical trees of Brazilian families.

In his book *What is Brazil* (2003), Da Matta makes pertinent points about our miscegenation, and most importantly for our reasoning: he claims that this mixture of races is one of the elements that allows us to say who the Brazilian is. In addition to this observation, he also states that prejudice is latent in Brazil when we see racial theories in which the races that make up Brazilian culture are classified hierarchically. He states that:

"Of course we can have a racial democracy in Brazil. But it will have to be founded on legal positivity that guarantees all Brazilians the foundation of all equality: the right to be equal before the law. Until this is discussed and practiced, we will always be using our mulattos and mestigos as a disguise for a social process marked by inequality". (Da Matta, 2003)

Da Matta's discourse on how miscegenation needs to be revised collaborates greatly with Amado's vision expressed through Arcanjo's actions. Da Matta (2003) states that it is necessary to unmask the real interpretations of the racial theories that marked academic discourse in Brazil for a long time. Such as that of Gobineau, who was French consul in Brazil during the imperial period. In his book *The Moral and Intellectual Diversity of Races* (1856), the problems go beyond the attempt to affirm the white race as superior. He also sought to condemn our multiracial people, certifying that the real end of Brazilians was miscegenation. These ideas were propagated with great intensity among academics, who in Amado's book are represented by the head of Bahia's medical school. Da Matta (2003) interprets this hybridity as one of the positive aspects of our culture, the mestigo being a kind of perfect synthesis of the best that can exist in black, white and Indian. In a more formal and academic attempt to bring greater visibility and value to his discourse, Arcanjo published a book on miscegenation in Bahia:

"As the pages and chapters unfolded, Pedro Arcanjo forgot teachers and theories, no longer interested in refuting them in a polemic of assertions for which he was not even prepared, but in narrating Bahian life, the miseries and wonders of this daily life of poverty and trust; to show the decision of the persecuted and punished people of Bahia to overcome everything and survive, preserving and expanding the goods of dance, song, metal, iron, wood, goods of culture and freedom received as inheritance in the slave quarters and quilombos." (Amado, 1969, p.164)

Arcanjo's great strategy, besides considering the physiognomic marks of miscegenation, was to observe which values and customs of "non-white" culture were present in everyday life. From the time of the book to the present day, there is no Brazilian city with more Afro-descendant cultural effervescence than Salvador. Considering these cultural values, expressed on a daily basis in people's way of life, it is important to realize that miscegenation has overflowed into physiognomies, it also defines people's way of life. Miscegenation, as Arcanjo defended and published, is an ontological element of his people's cultural bias.

As a counterpoint, Amado also wrote for later reflection how much the racial question and miscegenation itself also reflect the universe of social inequalities and spatial segregation. Color is an element of differentiation, but in other scalar references that go beyond Bahia, all Bahians are of the same racial origin: mestigagem. Zabela, a character representing a decaying aristocracy of European origin and a resident of Pelourinho, became a close friend of Arcanjo. She would be the writer's main source of research for several of his books, especially on the bourgeois families of Bahia. She makes an important point to Arcanjo in this future publication. Let's take a look at her statement:

"Whites? Master Pedro, don't give me any whiteness in Bahia. Don't make me laugh, I can't, the pain cuts me. How many times have I told you that pure white in Bahia is like sugar from a mill: all brown. That's true in the Reconcavo, let alone the Sertao." (Amado, 1969, p. 277)

Zabela, who came from Europe and was familiar with European Caucasian races that prided themselves on their purity, claims that this type of purity doesn't exist in Bahia, further confirming that Amado's text agrees with the ideas of Da Matta (2003), who states that miscegenation is a defining element of what it means to be Brazilian. However, Zabela does not dismiss the current and unquestionable racial prejudice that permeates the wealthier social classes in Brazil. She states that:

"[...] some say what they think: black and mulatto only in the slave quarters. Others declare themselves liberals, egalitarians, you see, and the lack of prejudice lasts until marriage is mentioned" (Amado, 1969, p. 277).

In his third book, Arcanjo, overwhelmed by indignation at the racial inequalities that had been generating conflicts where the limitations of the mestizos were latent in the face of the privileges of the nobler and "whitened" classes, decided to research the sources of mestigagem in Bahia, with the collaboration of Zabela. After much research, he came to the conclusion that "there was no family without mixed blood. [Pure white was non-existent in Bahia, all white blood had been enriched by indigenous and black blood, usually both. The mixture began with the shipwreck of Caramuru, never stopped, continues current and accelerated, and is the basis of nationality." (Amado, 1969, page: 323).

One of the most confrontational moments in the writer's book was the consequences generated by the genealogical theme published by Arcanjo. At this juncture, the noble families of Bahia, especially the figure of Nilo Argolo, reacted with great revolt to the publication. Nilo, Arcanjo's greatest ideological rival, a doctor at the Faculty of Medicine in Bahia and defender of segregationist racial theories, was one of the main targets of the publication. Pedro Arcanjo, "listed the noble families of Bahia and completed the genealogical trees, which in general paid little attention to certain grandparents, certain marriages, bastard and illegitimate children. Based on irrefutable evidence, there they were, from stem to branch, whites, blacks and indians, settlers, slaves and freedmen, warriors and scholars, priests and sorcerers, that national mixture." (Amado, 1969, p. 324).

Based on the miscegenation of the Brazilian people, with no proven class distinctions, Arcanjo - personifying Amado - manages to reveal two paradoxical feelings about the human being constructed in Tenda dos Milagres: that of Zabela, who treats miscegenation as something irrelevant to the definition of nobility in the people, since in her opinion there is no purity of race in the country and, as it were, in Bahia. And on the other hand, the provocative result of Arcanjo in the figure of Nilo

Argolo, a social actor who vehemently denied the existence of miscegenation in all social classes, having not only to accept this point proven by the protagonist in the publication of his book, but also to swallow the deconstruction of his racist ideas. This is perhaps one of the most politicized passages *in Tent of Miracles.*

3 - The city of Salvador from Tenda dos Milagres:

The city of Salvador in *Tenda dos Milagres* is based on a dualistic society: the rich, generally occupying the upper city beyond the Pelourinho; and the poor in the segregated urban space. Another of Amado's positions that occurs throughout the text is to define the "white" space of the city as opposed to the black space. In other novels, Amado often portrayed[10] retrograde ideas about the mixed-race population, but in *Tenda dos Milagres* there was a difference due to his use of location: the Faculty of Medicine in Bahia, a place of knowledge and at the same time a place where discrimination against mixed-race people was perpetuated. At the beginning of the 20th century, the Medical School of Bahia in the novel was portrayed as a niche of sub-literature, where academic work included studies of racist theories combined with anthropological studies.

The spatial segregation represented in *Tenda dos Milagres* was clear from the fact that the candomble terreiros were located in marginalized areas of the city. According to Sellman (2013), the character Pedrito Gordo, the police chief who persecuted the terreiros, was inspired by a real-life public figure, Pedro Azevedo Gordilho, who led the violent repression of Candomble in the 1920s in Salvador (page 238).

As in real life at the time, the authorities systematically banned any cultural manifestation of Salvador's black population, from afoxe to street candombles, mixed or black Carnival blocks that would go out in independent parades. In the novel, the Police Secretary had forbidden it:

"For ethnic and social reasons, in defense of families, customs, morals and public welfare, in the fight against crime, debauchery and disorder," the afoxes' exit and parade began in 1904" (Amado, 1969, p. 70).

The Afoxe filhos da Bahia block, led by Pedro Arcanjo, however, did not comply with the order and brought their invincible fighters to Rua Zumbi dos Palmares, an event mentioned above. Both in the novel and in real life, these acts by the authorities legitimized the racist theories of the Faculty of Medicine. They provided scientific paradigms to justify oppression. The conservative classes, for their part, saw the afoxes and their music as a disturbance to the established order:

"The gazettes protested against the "way in which the Carnival festival, that great festival of civilization, has been Africanized among us". [...] "The authorities should ban these batuques and candombles [...] chanting the abominable samba, because all this is incompatible with our state of civilization," cried *the Jornal de Noticias, a* powerful organ of the conservative classes." (Amado, 1969, p. 73)

---

[10] **Jubiaba (1935), Mar Morto (1936), Capitaes da Areia (1937).**

The counterpoint to the oppressive ideas of the Faculty of Medicine also worked miracles in the Pelourinho neighborhood. By making Tenda a university of popular knowledge, Amado defended the cultural manifestations of the poor as legitimate knowledge of the true heritage of the people of Salvador. In *Tenda dos Milagres,* "Pelourinho is a place of freedom that defends the human creation of freedom against an oppressive past". (Sellman, 2013).

Another illustrative and collaborative point for this research is Amado's construction of the terreiros in Pelourinho. When the writer presents some of these cultural manifestations in the historic district of Pelourinho, he elevates this place as a social and cultural model for the rest of the city of Salvador. This is how a new geographical imaginary is constructed, and with this bias, a new way of conceptualizing the city is inaugurated.

The territoriality of Pelourinho stretches through the streets of Salvador's city center, which are quite peculiar in their layout: winding, narrow hillsides such as Tabuao and Maciel; its hinterland such as Baixa dos Sapateiros and Largo da Se; the areas that surround it such as Portas do Carmo, Santo Antonio and Tororo and other areas beyond such as Rio Vermelho and Sete Portas. Pelourinho is a city within the city of Salvador, made up of its lower-class neighborhoods. Jorge Amado's Salvador was born in the streets of Pelourinho. Thus, the mixture that is observed through the pages of *Tent of Miracles*, with Pelourinho as the stage for the events, is what we will call the geography of Salvador.

There are many objects present in the Pelourinho landscape that further consolidate the argument made here: understanding the Amadian man from the perspective of his places of belonging. Unlike the Faculty of Medicine, Pelourinho brings together different ethnicities and cultures in the same geographical environment.

The Church of Nossa Senhora do Rosario dos Pretos, located in the Pelourinho district, also inaugurated syncretic services, mixing Catholic prayers with Candomble drumming. In the building next to the church there is a capoeira school. A combination of art, dance and fighting, capoeira originated with the black slaves who arrived from Africa. Throughout the book, Amado points out to his readers that black slaves used to be brought to Pelourinho for punishment. In the end, by concentrating them in the same territory, this contributed to the resistance of their native practices, and the capoeira school is a place of this kind. Furthermore, where capoeira was practiced, it was also said that a place of resistance to the oppression and segregation of the city was born there. Pelourinho was seen by the people of Salvador, as Amado represented it, as a *free territory.*

Sharing the leading role with Arcanjo, we have not one person, but a place of great human expression in the book: the Tenda dos Milagres itself. On its way through the "popular university" of Pelourinho, this Tent is accompanied by the presidents of its office.

The Tent of Miracles also represents a place in the city of mixtures. However, it won't be miscegenation, but a mixture of popular artistic references: art, literature, sculpture, music, capoeira, dance, theater, etc. These references also reflect the multiple uses of the city's historic center. It also functions as a popular printing house: the cordels of the street minstrels are printed and published there. It was also in the Tenda dos Milagres that Pedro Arcanjo printed his ethnographic studies. The Tenda is a space that accommodates both the popular and scholarly publications of Bahian culture.

Another very important point to consider about Tenda dos Milagres is the extent to which it illustrates Freyrian ideas of miscegenation throughout the narrative. The suggested conceptualization of the city according to the logic narrated by Amado is that Salvador is a mixed-race space. The positive aspects of Afro-Brazilian culture in *Tenda dos Milagres* come from popular practices as well as candomble.

Through Arcanjo's ideas, Amado claims that popular traditions and their knowledge are an integral part of Salvador's defining heritage. Furthermore, he uses these traditions to bring together the two extremes of the city, erasing the difference between erudite and popular culture: "one day the orixas will dango on the stages of the theaters" (Amado, 1969, p. 248).

In *Tenda dos Milagres,* Amado describes the city in a dialog with his youthful memories, delimiting what defines the urban space of Salvador.

4 - Mixed city, mixed country:

The conflict between the eugenic ideas of the Bahia Medical School and the miscegenation theories of Pedro Arcanjo represented a battle for Salvador's identity. Within this perspective of analysis, Sellman (2013) develops a proposal that will contribute directly to our definition and understanding of the geographical foundation of being and the founding element of place. According to Sellman, there are documents dating back to the 16th century in which the miscegenation of the races was already a fundamental element in the formation of the population. And, going further, places can be considered criteria for understanding ethnicities. In this context, when Arcanjo defends miscegenation, he also protects the defense of difference, in opposition to white European culture as an element of segregation for blacks. And we can say that this racial segregation is also spatial, since it is shown through the city's neighborhoods, i.e. territorial cutouts, where we associate the black or white identity embedded in the inhabitants of the city of Salvador.

Miscegenation in the Amadian novel is not just a mixture of races, but also a mixture of places and cultures. At the beginning of *Tenda dos Milagres,* Pelourinho is presented as a neighborhood with a multitude of stores, ateliers and various skills. In addition, Tenda dos Milagres, the popular university of Arcanjo and Lidio Corro, becomes a metonymy for Pelourinho by displaying a myriad of fungi.

The idea of the poor mestigo and the black population having access to information was considered quite dangerous by the elites in the early decades of Brazil's Republic, when this part of the narrative takes place. According to Sellman, Arcanjo's ideas, which initially appear in the book, that "the face of the Brazilian people is mestiza and its culture is mestiza" (Amado, 1969, p. 125), concatenate into a subversive reading: a lower-class man uses an instrument of mobilization, written language, to propagate his ideological beliefs. In a later study, *Apontamentos sobre a mestigagem nas famílias,* Arcanjo makes an even more profound argument:

"Pure white was non-existent in Bahia, all white blood was enriched with indigenous and black blood, usually both. The mixture began with the shipwreck of Caramuru, it has never stopped, it continues to flow and accelerate, and it is the basis of nationality." (Amado, 1969, p. 258)

Amado, by mentioning the founder of the original syncretism of the city that later became Salvador - the person known as Caramuru[11] -, also suggested that the miscegenation of Brazilian identity began in this city. To prove his point, Amado has Arcanjo research the list of genealogical trees of the main noble families of Bahia, and includes an extensive list of "politicians, writers, journalists, and even barons of the Empire, diplomats and bishops, all mulattoes, the best of the country's intelligentsia" (Amado, 1969, p. 258).

In another vein, miscegenation and the idea of mestizo popular culture are also dealt with in *Tenda dos Milagres* from the perspective of the terreiros, such as the *Terreiro de Jesus.* And an interesting clash in the book appears through dialogues between Pedro Arcanjo and Professor Fraga Neto. This dialogue will also represent the two matrices of thought that seem paradoxical in Amado: rationality and mysticism. Fraga Neto, a progressive professor at the Faculty of Medicine, asks Arcanjo why he continues to go to the terreiro, since he has now become a man of reason. Arcanjo replies that it is a cultural and political choice, not a faith choice, arguing:

"Terreiro de Jesus, everything mixed up in Bahia, professor. O adro de Jesus, o terreiro de Oxala, Terreiro de Jesus. I'm a mixture of ragas and men, I'm a mulatto, a Brazilian. Tomorrow will be as you say and wish, it certainly will be, man moves forward." (Amado, 1969, p. 247)

The *Terreiro de Jesus* is a place that has many meanings for many cultures. The term 'terreiro' is also used to refer to Afro-Brazilian Candomble temples. By defending the terreiros and attending Candomble ceremonies, Arcanjo carnavalizes the city's urban space, calling for greater dialogue and acceptance of the country's formative cultures. These cultures do not represent *Terreiro de Jesus* individually, they are also immersed in the hybridity of its formation.

Arcanjo/Amado suggests how conflicting cultures can meet and coexist in the same geographical environment, enriching the geography of the place. Amado's Geography defines his man by listing these cultural elements of miscegenation.

Furthermore, Arcanjo seems to be proposing that Salvador is the vanguard city of Brazil, the place where the ideal of a Brazilian identity has already been realized.

Another interesting angle on miscegenation in Amado's work is how optimistically he views this cultural value of Brazilian identity. It is even through a mulatto character who has "risen in life" that Amado reveals who he believes the true Brazilian miscegenation to be. He approaches the question with the following proposition: be mulatto and mixed. What matters is that you want to study, so he believed that the mestigos would have the possibility of social ascension. This is how he portrays Tadeu Canhoto, the example of his true idealized man in *Tenda dos Milagres.*

Tadeu Canhoto is the bastard son of Arcanjo and the mysterious Doroteia, who is supposedly an 'iaba', a female demon in the world of the orixas[11] Under Arcanjo's tutelage, Tadeu finishes his studies and graduates from the Polytechnic School. The character's departure from Salvador is portrayed ambiguously.

With the help of one of the professors at the Polytechnic School, Tadeu moved to Rio, "where he joined the team of engineers who, under the command of Paulo de Frontin, were transforming the country's capital into a modern city" (Amado, 1969, p. 98).

Referring back to the historical context, in real life, Frontin was the engineer who, under Mayor Pereira Passos, led a radical urban plan to modernize Rio de Janeiro in the 1900s. In Rio, Tadeu became one of Frontin's favorites:

"Paulo de Frontin solved nothing, no detail of the great urban plans, without listening to his opinion, appointing him responsible for the most difficult tasks. In practice, Tadeu was building the new Rio de Janeiro" (Amado, 1969, p. 227).

Thus, from the perspective of work, what defines Tadeu Canhoto is no longer his color, but the job he occupies. And, with this turn of events, his existence as a mulatto from Bahia who frequented *the Tenda dos Milagres* is transformed into a well-accomplished engineer living in Rio de Janeiro. Amado leads us to another very reflective reading about miscegenation as the founding element of man in Salvador: this mulatto who is the character Tadeu Canhoto, in order to be seen as a man beyond his color, needs to study, work and leave the city. Otherwise, he would remain in the same social condition.

Another contradiction is also born in Tadeu Canhoto when he moves to Rio de Janeiro. In another city, experiencing other geographies, he is no longer the mulatto from Bahia. What's more, he still disseminates work that supports segregation permeated by the argument against miscegenation. The works carried out under the coordination of Pereira Passos, led by Paulo de

[11] Iaba, Yaba or Iyaba, meaning Queen Mother, is the term given to the female orishas Yemanja and Oxum, but in Brazil this term is used to define all female orishas in general instead of the term Obirinxa.

Frontin, which were the front line of Tadeu Canhoto's work, had a very clear political role in the city of Rio: to 'clean up' the city center. According to the city's hygienist policy at the time - a true fact that inspired the Amadian narrative - a good city was a clean city. And, contradictory to Canhoto's life story, the poor population of Rio de Janeiro, to which he now represented the opposition, as in Salvador, was mainly black or mestizo. The same ideas that inspired the Faculty of Medicine in Salvador also guided the transformation of urbanity in Rio de Janeiro (Sellman, 2013).

Tadeu Canhoto, as the novel portrays, contradictory to his life story, became an important figure in the segregation process: "one of the engineers responsible for the urbanization of Rio de Janeiro" (Amado, 1969, p. 257).

When Tadeu Canhoto went to Rio de Janeiro, he lost contact with Arcanjo and only saw him again one last time at the end of Tenda dos Milagres. Understanding the complexity of the man who reveals the character Tadeu becomes a major challenge within the debate on miscegenation in Amado's work. He offers a critique of the official discourse on miscegenation and national identity at that time. On the other hand, Tadeu's story, according to Arcanjo's logic, is not an example of the miscegenation of liberation, but of the reinforcement of a city like Salvador that is spatially organized according to a logic of binary segregation: white territorialities and geographies and mixed-race and/or black territorialities and geographies.

Focusing on this geographical environment, which emerges from this man who is being problematized here as a result of miscegenation, and this in turn as a conditional and existential element of it, some observations can be made. The resulting landscape remains the same at both ends of Tenda's historical spectrum: the social division in the city, which is also spatial - the poor population is still segregated in the city's historic center and outlying neighborhoods - while their uses of the environment are restricted or suppressed.

Analysing the spatio-temporal order of the geographies that exist in *Tenda dos Milagres,* the established power only allows for the spread of a sanitized version of Arcanjo's ideal city: the hybrid city of multiple uses and the mixture of hegemonic and popular cultures is transformed into a folkloric take on the existing order. The end of the novel seems to point towards an existential utopia for the city of law defended by Arcanjo.

5 - Jorge Amado and mestizo geography:

In an interview with *Caderno de Literatura Brasileira* (1997) in his honor, Amado said that "syncretism is characteristic of Brazil. There is this mixture here, so we can't stop thinking about it in order to think about this affirmation of differences." (page 55).

For Amado, the most striking elements of Brazilian identity are undoubtedly the mestigagem, the mixture. We are not this or that, we are everything: white, black, Indian. This is what makes us unique and gives us real importance.

The blacks in Amado's work are those who struggle to see themselves respected by the recognition of another culture that is not the dominant white one that has so far represented the idea of nationhood. Or at least to have this miscegenation recognized, which cannot be denied as an element that identifies us. For Assis Duarte (2012), in *Tenda dos Milagres* there is, in parallel to the discourse on the elevation of the black race, a praise for miscegenation and the Brazilian cultural melting pot, to a certain extent tributary to Gilberto Freyre's theses on racial democracy.

If you read the novelist's main texts, you'll notice that they embody the main polarities of social life. Or perhaps the dualities inherent in the human condition? First of all, we would point to the clash between nature and culture. Then, the inevitably asymmetrical situation between men and women. Then the contrast between the countryside and the city, or the confrontation between land and sea. These are thematic universes expertly dealt with in Jorge Amado's fiction. In his critical view of Brazil, Jorge Amado works with the historical contrast between white and black and offers, as a positive response, the immense estuary of miscegenation. And he mitigates the different opposites mentioned above with the force of progress, the cunning of love, the joy of living, the breaking of rules and religious syncretism. The latter are better demarcated as geography in *Tent of Miracles.* The human being in Amado's work is one in whom the dualities of his being dwell in his essence.

Tenda dos Milagres addresses fundamental problems in Brazilian society. In this narrative, the reader is confronted with an unprecedented fact in Brazilian fiction: the superiority of black people. In an interview in Caderno de Literatura (1997), Amado states:

> "Perhaps Pedro Arcanjo is the most complete of all my characters. [...] When asked how he managed to be materialistic and at the same time exercise his functions in candomble, he replied: 'my materialism doesn't limit me.

It is possible to think about a geography in Jorge Amado's work based on some of the characteristics that the author developed in his characters and in the setting of his plot: in other words, in the geographies constructed in his environment, in the social web presented in the relationship between this environment and his characters. This relationship is fundamental to the context of the plot. It is important to emphasize that, in the case of a literary career such as Jorge Amado's, which is dense, extensive and spans very different periods in the author's career, any of its characteristics must be seen without looking at the whole picture. However, the plurality of themes to talk about the human being are very expressive in his work.

The theme of miscegenation in *Tent of Miracles* was made all the more surprising by the naturalized way in which it was portrayed: often, the naturalized sense of social fact confused us into believing that it really was a fact of nature and not of culture. For Amado, it was important to affirm that the great hero of his story could only be the mulatto, the black woman with a hard life, because only the fact that she was able to make a living with dignity and joy, would make the members of this social scene that was Bahia of the ghettos and poverty, of a tense geography and immersed in a reality of urban restrictions, a fact of great admiration. Amado's heroism is the ability of his men to find the will to live in an environment especially full of adversity.

With regard to the main theme of geography in *Tenda dos Milagres,* thinking about ethnicity would go beyond the racial issues addressed and the clear differences between blacks and whites in the four corners of the world. To think about these issues is for Amado to make the Brazilian people look in the mirror, and see a Brazil far beyond the South and its proposed concepts of what we would finally call the nagao.

Amado's black man is the one who lives loose on the hillsides of Bahia, who puts his friends first, who parties, dances, drinks cachaga, makes children indiscriminately and profusely with mulatto, black, Swedish or Finnish women. The black man whose affinity with African streets and roots is greater than his repressive elements. This black man, by mixing with the white man, gives birth to the Amadian being, the Bahian being and why not the being of universal characteristics.

By addressing the affective social intersection of different races in his works, the author raises a crucial issue in his utopia, in which social transformation will have its roots in the dynamics of popular culture. The Bahian novelist starts from the fact that we are undeniably ethnically mixed, and then discerns cultural mixing as an essential feature of our identity. This is to emphasize how plural a country with elements like Brazil is. The recognition of mestigagem and the celebration of cultural richness in his work, by being constituent elements of geographies, can be celebrated as theories that help us understand Brazil where this Amadian symbolic universe, based on utopias and concreteness, can be the light in the construction of a political project that would not need, for example, doctrinaire political thoughts. It's the people in their midst, speaking of themselves and for themselves.

Another important consideration is that extracting this geographical content from the works listed here also necessarily involves understanding a spatial reading of the city of Salvador. Even in different historical contexts. It may be that today there is a sense of a Salvador that no longer exists, that is nostalgic. We know about the different processes that line up the construction of the urban space of large cities: the process of peripheralization and gentrification that culminate in deeper spatial segregation. Pelourinho, once a place of identity for vagrants and a decadent class, as Jorge Amado often addressed it, albeit in an amusing way, is now one of the city's centers of tourist functionality. Pelourinho now welcomes tourists and has taken on a new dimension.

At a time like today, when the city is very much thought of as a market, it is not enough to sustain only this bias towards geography. The city-habitat is also the city that is inhabited, that has habits, what in popular language we will call the pulsating life of people in the city of Salvador.

6 - The geography of the city of Salvador in Tent of Miracles:

The city is a place of exchange, a symbol capable of expressing the tension between the geographical rationality of its objects and the tangle of its human existence. Even when demolition or construction, generated by the typical movement of "progress", end up erasing the urban memory swallowed up in the city's geographical environment, it is possible to recover these geographies through books and literature.

Certainly, the city has been one of the most present themes in the history of literature and everything that derives from it. In *Tenda dos Milagres,* the story of the city of Salvador shows us a hybrid city of colors, habits and people: it is the fruit of miscegenation. And, in this respect, Jorge Amado is intimate with the city of Salvador. Few cities have the grace, description and realism of the writer's lines.

The Salvador of this book is that of the late 1960s, a city that was the result of concentrated, abrupt and exclusionary urbanization. Bittencourt Andrade (2004) briefly describes the historical occupation of the urban site of the city of Salvador at the time of *Tenda dos Milagres:*

"The Atlantic shore was formally occupied by the elite (except for a few but densely populated invasions by the poor), bordered by Paralela avenue forming a conurbation up to Lauro de Freitas. The 'core' of the city received immigrants with little qualifications and money, resulting in low-income and disorderly occupation;[....] The lower city, following the so-called railroad suburb, lives a hybrid reality, with middle and lower class residential neighborhoods, which have historically occupied that region and resist change, and areas of intense poverty."

A geographical understanding of a previous space can be gained from reading the books and their respective geographies that we will identify in their pages. In *Tenda dos Milagres,* the occupation of this urban site in Salvador, as described above, is not so linear and continuous. However, the clash between classes in the hybrid spatial reality enshrined there is clearly perceptible. Thus, the complex urban and social fabric of Salvador oscillates between the eugenicist intellectual preponderance of a minority in search of more power, space, status and money; and the struggle for survival and freedom to worship their values and identity, inherited from different races, by the vast majority. This social duality is a constant throughout the plot, and it is from it that we draw these geographies. This is where the geographical content emerges.

Chapter 3- Man and his land in Jubiaba:

Jubiaba, a novel written by Jorge Amado in 1935, portrays the journey of the character Antonio Balduino: from a boy from Morro do Capa Negro to a striking leader of trade union struggles in the city of Salvador, Bahia. Published when Jorge Amado was 23, *Jubiaba* was the first of his works to achieve critical and public success. The central intention of the work, apart from its novelistic view of popular life, is to suggest the protagonist's slow maturation towards political awareness. With some sensual and spicy ingredients from the Bahian scene, it is a novel characteristic of "socialist realism".[12]

The novel is called *Jubiaba, the* name of the people's saint father and spiritual guide, not the protagonist Antonio Balduino. Although the character who gives the novel its title, a Yoruba healer and priest, is remembered throughout the book, it is the black Antonio Balduino who dominates the scene. The saint's father represents the entire community of Morro do Capa Negro and its surroundings, and is the epic, religious hero who gives the people their identity throughout the narrative. In the course of the novel, the father of the saint is an insistent reference point for the black man Balduino, as a role model and life guide. A homeless man, Balduino becomes a successful boxer, but only for a while, because when he doesn't take his first defeat well, he leaves the ring and saves his life. Balduino continues along his multifaceted path with one constant: following Jubiaba's lead. The character only frees himself from this domination when Jubiaba is unable to explain the meaning of the strike, a meaning Balduino discovers at the end of the narrative, which will make him a new man. Of all the novels analyzed here, *Jubiaba* is perhaps the one in which the proposal of ontological experience and the discussion of geography is most explicit in a context of movement, of leaving Salvador to discover who he really is and returning to it.

The narrative is made up of tales and legends that are part of the Northeastern people's imagination, as well as stories about traveling sailors who live a life of constant danger. These stories educate the character Balduino, who builds his identity on the basis of these experiences, which become ontological in nature. The world to which Balduino belongs is filled with stories of his people, both his ancestors and the men who lived there: saveiristas, jagungos, sertanejos, workers, who are made known to him through the narrative voices of older people who were important in his formation: father Jubiaba, Ze Camarao, Mestre Manuel, aunt Luisa.

The Morro do Capa Negro plays an important role in defining who Antonio Balduino will be. Firstly, because of the landscape it reveals about the city, which often leads Baldo to moments of reflection about who he is at that moment. Then there are the people who live there. All the inhabitants of the hill end up becoming narrators of legends and stories. They sit outside their shacks every night to tell and listen to them. They are sad stories of abandoned women, of beatings well given, of deaths

[12] In Brazil at the beginning of the 20th century, in the midst of the debate about the formation of the Republic and the nation, the social novel emerged, characterized by its criticism of the ruling classes. Socialist realism was the official artistic style of the Soviet Union between the 1930s and 1960s. As Jorge Amado developed closer ties with the Communist Party of Brazil, his literature became more socially critical and took on a more political stance, as expressed in Jubiaba.

in defense of honor. And all this everyday life will be responsible for defining who the protagonist is - or who he becomes - throughout the novel.

In *Jubiaba*, the delimitation of the geographical environment contributes to the characterization of critical realism, in other words, the delimitation of the social issues of exploitation in a spatial logic. We see the space of vagrancy, the work of dockers, the harvesting of tobacco, work in the circus, etc. These spaces mark an era, a moment in the history of a people, as well as in the existential trajectory of the protagonist. In *Jubiaba*, there is a predominance of open spaces that configure the freedom characteristic of the character Balduino.

The context in which Amado wrote *Jubiaba* was a historical period of political and ideological radicalization as well as cultural activity. His socialist-realist writing reveals his Marxist position at the time and also his desire to denounce the social problems arising from capitalism in the country by looking at Bahia. Thus, *Jubiaba* can also be seen as a novel that places Amado's narrative, which is so regional, on a broader scale of the universal category. Since the social questions raised by the writer in the book are not exclusive to Brazil.

1 - Antonio Balduino: displacements and transformations

Antonio Balduino, despite not being the title of the novel *Jubiaba* as we have already heard, is undoubtedly the protagonist of the narrative. And the great theme that surrounds his entire trajectory is that his constant and continuous transformations translate more than his individual and subjective aspects: they bring meaning to the entire geography contained in this plot. His transformations take place in spatio-temporal circumstances in which his path is aimed at the geographic foundations contained in the work. Take a look at the following passage:

"Se Square had been flooded that night. The men squirmed on the benches, sweating, their eyes drawn to the platform where the black Antonio Balduino was fighting Ergin, the German. The shadow of the century-old church stretched over the men. Few lamps illuminated the platform. Soldiers, dockers, students, workers, men wearing only shirts and trousers, eagerly followed the fight. Blacks, whites and mulattos were cheering for Antonio Balduino, who had already knocked down his opponent twice." (Amado, 1935, p. 9)

This is the description of the first event in the book. We don't yet know the context of the plot, but the presence of the only character who will accompany the whole story and is still to be delineated is there: Antonio Balduino. The setting and the action are present, configuring geography and temporality. The classes are in place, even the racial conflict, all of which are themes that will be problematized throughout the book, and the figure of Baldo in the struggle does not go unnoticed either. Here in this first moment, the boxing match. Throughout the book, the fight will not only be of one type or form, but the lasting defining action of various circumstances in Antonio Balduino's life. Including occasions when his own struggles represented the foundation of the city of Salvador or Bahia itself. In other words, Brazil itself.

Details in *Jubiaba* also reflect Amado's political life at the time. For example, one of the aims of the Alianga Nacional Libertadora, in which Amado participated, was to fight against the influence of Nazi-fascism in Brazil. Consequently, *Jubiaba* begins with its main character, Antonio Balduino, defeating a white German in a boxing match.

Balduino's characteristics are very peculiar and chameleon-like: he is a man who is free in his unconsciousness and who naturally rejects the normative paths in which he would probably fit. Balduino has had the most diverse life experiences: street kid, rascal, samba composer, boxer, tobacco plantation worker, circus performer and dockworker. Amado's choice of starting with the boxing match could perfectly well have been replaced by any of the others, because in fact, the important thing to absorb as the narrative unfolds is the diverse path that Balduino takes to define what kind of man he will become.

Willing and quarrelsome, supportive and friendly, Balduino is searching for something undefined that he is fomenting within himself, a vague claim that torments him and a desire for affirmation that shows his "true path":

"What will Antonio Balduino, a little black boy who is only fifteen years old and already emperor of the black city of Bahia, lack? He doesn't know, nor does anyone else. But there is something missing, which he will have to cross the sea to find, or wait for the sea to bring it to him in the bowels of a transatlantic, or in the hold of a ship, or even attached to the body of a shipwreck." (Amado, 1935, p. 65)

The novel *Jubiaba* signals and endorses, according to the trajectory of its protagonist, the possibility of the dispossessed breaking away from the prevailing historical and social conjuncture. From this look at the empirical analysis of the work, we are explicitly faced with the realist-socialist profile chosen by Amado. Realism will come through the sense of struggle(s) that the protagonist will experience. The sense of struggle will mark much more than the character's first contact with the reader. Balduino emerges as someone who beats and wins, but who also suffers setbacks. His victory does not come without the sacrifice of punches to the face, which already points to the heroic facet with which the text will cover his figure. His reflections on his struggles and, above all, on who he is define him as he lives.

The great enthusiast and critic of Jorge Amado's work, Bastide (1972), pointed out that a brief study of what a realistic novel would be like:

"[...] the first novels by Jose Lins do Rego and Jorge Amado, who are the two great writers of the new regionalist school in the Northeast, merely continue the old naturalist school prior to modernism, which pretends, after Emile Zola and his 'experimental novel', to be the faithful painting of a certain sociological milieu, a 'slice of life', a document as scientific and exact, or perhaps more so, as that which could be presented by a specialist in social sciences interested in the same problems."

The message that Amado wants to convey is that the novel should also have the role of denouncing the social ills inscribed in the geographical environment portrayed. And dealing with these complications in Bahia undoubtedly touches on racial issues. Therefore, he emphasizes this message of a social realist novel through the creation of a main character who is black, and a member of the poorest and most oppressed group of the Brazilian population, who becomes aware of the issues surrounding him.

Jubiaba is considered one of Jorge Amado's masterpieces. Firstly, because in this book the black Antonio Balduino plays a fundamental solo role in the narrative, as he also leads the secondary characters on their journey. These: soldiers, porters, workers, blacks, mulattos, even whites who gain prominence, are homogenized by the same conditions of poverty. Secondly, because of the general theme of the work itself, where the writer has managed to harmoniously merge sociological documents, revolutionary demands and poetry into a single unit.

Antonio Balduino is on a quest for his own salvation, which is also the rescue of his people's hope. Moving from the African religion of the candombles to an awareness of the class struggle, he becomes the leader of himself and those around him. This succession of two moments in the hero's life, which we could call negritude and Marxism, is a great expression of the syncretism that accompanied Amado's ideas throughout his life.

Amado built a character in Balduino where his external experiences, with the other and with the environment, easily help us to understand the formation of his existence. In this constructive and dialectical process, we see his geography being born.

2 - Balduino's black protagonism:

For some, *Jubiaba* inaugurated the appearance of the first black hero in Brazilian literature. Afro-Brazilian elements will recur in Antonio Balduino's almost epic journey: candomble, the terreiro, the father of saint Counsellor, the atabaques, capoeira, the samba circles, the mungunza made by his Aunt Luiza, the Morro do Capa Negro. And it's worth pointing out that these elements are not merely aesthetic treatments in the work, but the founding nature of the man Balduino and, in turn, of the entire novel. Amado places them in the book through the thoughts of his protagonist:

"Blacks are still slaves and whites too. - said a thin man working on the pier. - Every poor person is still a slave. Slavery isn't over yet... The blacks, the mulattos and the whites all put their heads down. Only Antonio Balduino kept his head up. He wasn't going to be a slave." (Amado, 1935, page 36)

Balduino is not the first black main character in Brazilian literature. However, he is the first black protagonist in the country's adventure fiction genre: a larger-than-life character who is capable of both tenderness and extreme violence. *Jubiaba* presents the process of his formation - or education - from childhood to adulthood, through various picaresque adventures in and around the city of Salvador. Thus, the unprecedented nature of a novel whose hero is black, poor and from the slums could not be overlooked. Add to this the fact that he earns his living from working in the fields of tobacco or on the docks as a stevedore. Thus, "Jorge Amado knew how to valorize the black man, lending him a heroic tessitura and extracting convincing dramatic substance from him. He presents the black man not only in his social measure, but in his intimate concerns" (Portella, 1972).

The black Balduino is clearly a transformative agent in Salvador, an Exu like figure who is programmed to disrupt the city's precarious status quo. Through change, Balduino/Exu is helping to found a new order. When this happens, he is the messenger of both the gods and the proletarian movement: he transforms chaos into effective change. Within these questions that permeate Antonio Balduino's blackness, *Jubiaba* is a novel that is very representative of the author's alignment with the prevailing tone of racial studies at the time of his writing. Amado's black is a definition that goes beyond a mere phenotype. It is also a social positioning. Rossi (2009) puts it this way:

"Black people and their descendants appear in Amado's eyes as the oppressed par excellence, because they have not yet managed to realize their 'true' abolition, doubly excluded: сото гада and сото class. It's no coincidence that hill and city are well-defined spaces in Jubiaba. If, at first, the slum provided the keys for Balduino to understand his oppression as a black man, then his move to the city allowed him to perceive the broader shackles that enslaved him."

Jorge Amado found in Afro-Brazilian culture a beauty and a theme worthy of interpretation, while at the same time coating it with a heroic sense, based on what it contained from a group that had historically been enslaved and placed on the margins of society. The author's problematization of black people reveals a very clear ontological reflection: identification and the feeling of belonging or not belonging to a race are defined by the specific positions occupied by the subjects in the social structure and in the field of political struggles. In *Jubiaba,* "Jorge Amado seems to work with the idea that one is not black, but one is black" (Rossi, 2009).

The powerful racial issue in *Jubiaba is* explained throughout the novel in Marxist terms: it is a class issue. By reducing the racial question to material terms, Amado presented communism as an umbrella solution for all social shortcomings in Brazil. Amado himself explained this point of view many years later to his French translator, Alice Raillard (1990):

In *Jubiaba,* the problem of raga is posed in such a violent way that, at the end of the book, Baldwin realizes that the problem of raga is first and foremost a class problem. The raga problem is not the cause, but the consequence of the class problem: the problem of the poor and the rich, the slave and the master (pp. 86/87).

Marxist themes permeate much of Jorge Amado's work, both in *Jubiaba* and in several of his other novels. This theme will be further problematized below. However, it is important to note here the relationship it has with the theme of race in this book. In his dealings with the black people - their customs, their beliefs, their speech - Jorge Amado gives free rein to his poetic inspiration. The form is almost always crude and primitive, just like the subject matter, but with a rhythm interspersed with Afro drumming. The Marxist Jorge Amado could sound contradictory when he is also a spiritualist. But it is there, in this syncretism, that Amado blends his fantasy with the reality of his people. It is easy to believe that, as a good Bahian, he believes in Iemanja, Oxossi; that he has faith in the fathers of saints with Jubiaba and their prayers in nago; that he believes in the miraculous power of herbs and in all the other orishas of candomble. In his writing, these are myths that the writer's intelligence does not try to repel; on the contrary, it welcomes them with pleasure, even accepting them as characters, active and present.

3 - Antonio Balduino's movements and the construction of his environment:

One of the main characteristics of Jorge Amado's novel, in addition to Balduino's black protagonist, is the role that the social space plays in the plot of the story. We return to the argument about the importance of the "setting", as discussed in previous chapters. We have a text in which the popular epic tradition, inserted in the social space, puts the protagonist's inner self in the background. In other words, Balduino is a protagonist who is defined much more by his outward movements than by his zones of introspection. His subject exists and is constructed from the geography of his external world.

His movements range from the character's childhood formation in Morro do Capa Negro to his blossoming as a proletarian leader on the docks of Bahia. The first movements of the narrative indicate that the protagonist Balduino has a public life, that he exposes himself in the streets in order to receive rejection or applause. Balduino is almost always in public: with the boys from the slums, leading street punks, exposing himself in boxing rings and circus rings, getting involved in fights at the fairs, charming women with his sambas. And later, as an adult, during the strike, he took part in picket lines, gave speeches at assemblies and took part in the ceremonies of the Jubiaba terreiro.

According to Assis Duarte (1996), there are seven movements of rupture and ontological reconstruction in Antonio Balduino. Each of them has a well-defined spatial location, with supports and situations that show the protagonist's process of existential construction in the outer movement of social space. The first of these moments is his childhood in Morro do Capa Negro, when he lived with his Aunt Luisa and, at the age of eight, was already leading a gang of kids in their mischief. *Jubiaba*'s text emphasizes his loose life, without strict family control. At dusk, however, the boy settles down before the city lights up:

> "Antonio Balduino would stand on top of the hill watching the row of lights that was the city below. Guitar sounds crept over the hill as soon as the moon appeared. Mournful songs were sung. Seu Lourenpo Espanhol's shop was full of men who came to talk and read the newspaper he bought for the pinga customers.[...] He roamed freely on the hill and still neither loved nor hated. He was as pure as an animal and his only law was his instincts. He rode down the slopes in a mad rush, rode broomstick horses, had little conversation but a broad smile. He soon led the other boys on the hill, even those much older than him. He was imaginative and had courage like no other" (Amado, 1935, p. 16).

Balduino's gaze is contemplative and at the same time inquisitive. He searches for meaning in the labyrinth of lights that come on in the streets and avenues. Amid the sounds, voices and shapes of bodies circulating in the distance, he searches for something still nebulous: his identity. We read:

"The desire to know (and dominate) the outer space is visible, correlating with the need to find one's own destiny. This search will explain the character's mobility, his continual throwing himself outwards as the means and motor of self-knowledge" (Duarte, 1996, p. 25).

His aunt and tutor Luisa falls ill, and this event becomes Antonio Balduino's first move out of Morro do Capa Negro. The protagonist is pushed out of his place of genuine belonging, knowing that he must return to it. He leaves the slum to be raised in Comendador Pereira's house, the second movement in his trajectory. The fate of returning to the slum is announced by the saint's father Jubiaba, also highlighting his first major moment in the narrative. His childhood ends with the saint's father saying goodbye: "when you grow up, come here, when you're a man" (Amado, 1935, p. 52).

Two passages are interesting in this second movement. The first has to do with Baldo's awareness of his change, and at the same time sets up his first shock of social reality:

"Antonio Balduino was amazed at the size of the house. He had never seen anything like it. In Morro do Capa Negro the houses were small, made of beaten clay, with box doors and zinc roofs. They had only two rooms: the dining room and the sleeping area. But not the Commandador's sobrado. How big it was, how many rooms it had, some even closed, a guest room always furnished waiting for someone who never came, huge living rooms, a beautiful kitchen.[...]Only then did the little black man understand that he was separated from the slum, that they had torn him away from the place where he had been born and raised, where he had learned so much, and that they had thrown him, the freest of the slum kids, into the house of a gentleman." (Amado, 1935, pp. 43/46)

The second passage marks the relationship between Antonio Balduino and Jubiaba. It is the saint's father who is responsible for defining his trajectory by prophesying his return. About Jubiaba and his relationship with Balduino in the context of his childhood:

"Antonio Balduino was madly afraid of Jubiaba. He'd hide behind the door and through the crack he'd spy the sorcerer coming, his little white cap, his curved, dry body, leaning on a stick, walking slowly. The men stopped to say hello. [...] Jubiaba always carried a bunch of leaves that the wind shook and muttered words in Nago. He would come down the street talking to himself, waving, dragging his old cashmere trousers on top of which his embroidered sweater would offer itself to the whim of the wind like a flag. On certain nights, strange sounds of strange music came from Jubiaba's house. Antonio Balduino stirred in his mat, he became restless, it seemed that the music was calling him. Batuque, the sounds of dangas, different and mysterious voices...[...]Antonio Balduino didn't sleep at night. In his healthy and free childhood, Jubiaba was the mystery." (Amado, 1935, page 18)

Balduino's childhood represented a lake of strangeness about Jubiaba. Much more so when observing the experience of the spiritual priest on the hill than through contact. At the time, Jubiaba was something mythical and mysterious to the boy who was Antonio Balduino.

From the age of 12 to 15, Balduino lived in the Comendador's house on Travessa Zumbi dos Palmares. He did small jobs, went to school and learned to dissimulate and lie when necessary. A servant and errand boy, he enjoys the gentle company of Lindinalva, and from there arises the platonical love that accompanies him throughout the novel and which will also be the protagonist of one of the most constructive chapters of his being. The conflict arises when the cook Amelia intrigues about Balduino's conduct with Lindinalva. Thus, "female references are lost, innocence is lost; the child dies and the hero moves towards a new persona. Grown up, he leaves and goes to live in the winding alleys of the city." (Assis Duarte, 1996 page 81).

The space of the street inaugurates the period of naughty freedom, in a sense a return to the mischief of childhood: Balduino leads the gang of teenagers who live on handouts and petty crime, setting up the third movement that delineates his being. Emperor of the streets, the city of Salvador becomes his kingdom:

"Antonio Balduino was now free in the religious city of Bahia de Todos os Santos and the saint's father Jubiaba. He was living the great adventure of freedom. His home was the whole city, his job was to run it. The son of the poor slum, today he owns the city." (Amado, 1935, p. 52)

During this period of collective wandering through the streets of Salvador, laughter was prevalent in the nooks and crannies of the city in *Jubiaba*'s narrative:

"Those were the good years, the free years, when he and his group dominated the city, begging in the streets, fighting in the alleys, sleeping on the quays. The group was united and the boys perhaps held each other in high esteem [...] They were united, yes. When one fought, everyone fought. And everything they earned was shared fraternally. They had their own love and loved the fame of the group.[...] They lived the same loose life for two years. Two years of running around the city, watching soccer matches and boxing matches, fighting, going to the Olimpia Cinema, listening to the stories that Gordo told without noticing that they were growing up, becoming men, and that the song that spoke of seven little blind men was no longer useful to them who were already strong, huge black men, knocking down mulattos on the pier, roaming the religious city of Bahia" (Amado, 1935, p. 70).

However, the tears of marginal life come with the death of one of the boys when he is run over by a car. The catastrophe marks the end of the period and the beginning of its fourth movement: Balduino returns to his place of origin, Morro do Capa Negro, as his father Jubiaba had prophesied. He's no longer the boy who left; he's grown up and adheres to his father's naughty legacy, perfecting his guitar and capoeira skills. He's already the seductive Baldo who knocks down the mulattos on the edge of the beach. This is how Amado wrote about this moment in his protagonist's life and the fate of the other boys in the gang:

"Only Viriato, o Anao, who was getting smaller and curvier, remained begging. The others were distributed around the city in various trades, factory workers, street workers, dock porters. Gordo went to sell newspapers because he had a good voice. Antonio Balduino went back to the hill of Capa Negro, and got into mischief with Ze Camarao, playing capoeira, playing the guitar at parties, going to Jubiaba's macumbas" (Amado, 1935, p. 72).

Music and capoeira frame the scene of youthful adventure: the hero gets involved in flings and fights and so his phase of existential rebellion is portrayed. His "lumpen" phase is also characterized[13] : where work is seen as punishment and unnecessary. This aforementioned rebellion, which is now latent, even if still without purpose, will begin to germinate as something consistent against the order that Amado criticizes so much in his realist/socialist appeal. Thus, "the refusal of work is inscribed as an affirmation of freedom, in a novelistic perspective of the idealization of vagrancy" (Assis Duarte, 1996).

In this phase of Balduino's pilgrimage, two spatial references are important for the construction of his life and his respective geographies: the Lanterna dos Afogados bar and the Terreiro de Jubiaba. The terreiro is the space of spiritual gratitude for life and also the meeting point for blacks and the poor to celebrate on the hill. It's where the purpose of their lives is sealed by faith. *The Lanterna dos Afogados* bar stands out as a space for celebrations and meetings of this bohemian lumpesinato.

The *Lanterna dos Afogados (Drowned Man's Lantern*) bar, an old black room in a colonial townhouse, was home to blacks who worked on the docks, sailors, women with easy lives, among others. Everyone used to go there to listen to music, fashions played on the guitar and heart-warming stories. The botequim belonged to the widow of a sailor who had set it up many years before. She was a dark mulatto who made rice pudding for the customers and "boia" for the sailors. When Antonio bought it, he renamed it and did a general clean-up of the place, but the customers didn't show up because of the sudden and unexpected change. The day after the change, the superstitious Antonio returned the sign with the old name and called the widow he had dismissed when he bought the bar. She went back to making rice pudding for the customers and booze for the sailors and, above all, returned to the double bed that had been occupied by the previous owner. Antonio Balduino's favorite hangout, this was the place where he drank, smoked and chatted with friends. It was also the place where Balduino took cabrochas to show the sailors and other regulars his virility.

Balduino embodies the experience of roguery and vagrancy, which is further emphasized by his return to Morro do Capa Negro. Between composing a samba and an ABC, his wanderings around the city lead him to frequent the typical lumpesinato hangouts such as the *Lanterna dos Afogados*. With his daily life on the port quays, the poor parties that took place there with samba and cachaga, the *Lanterna dos Afogados* became a place of experience for Balduino. For him as well as for the people of Morro do Capa Negro, it was a stage for meanings of existence and resistance. From this perspective, Balduino's concerns become disturbing and never leave him. The space-time of the

---

[13] The term lumpemproletariat (from the German Lumpenproletariat: 'degraded and despicable section of the proletariat', from lump 'despicable person' and lumpen 'rag, rag' + proletariat 'proletariat') or lumpesinato, or even subproletariat designates, in Marxist vocabulary, the population situated socially below the proletariat, from the point of view of living and working conditions, made up of unorganized miserable fractions of the proletariat, not only deprived of economic resources, but also devoid of political and class consciousness, and therefore susceptible to serving the interests of the bourgeoisie. The term, which can be translated literally as "rag man", was introduced by Karl Marx and Friedrich Engels in The German Ideology (1845). However, in Jorge Amado's text, this concept is romanticized by the writer: the idea of lumpen passes through the subject who wants to live without having to work, from the adventures of the streets.

Lanterna dos Afogados bar is perfectly portrayed in *Jubiaba:*

"The customers returned to the Drowning Man's Lantern. There they discussed long cruises with blond and black sailors. Schooner masters talked about the Reconcavo fairs where they would take their boats full of fruit. They played guitars, sang sambas and told heart-warming stories during the starry nights. And women came down from Ladeira do Taboao to Lanterna dos Afogados." (Amado, 1935, page 74)

Thus, the *Lan* bar *has that of the Drowned*, another of the spaces outside of Balduino but perfectly designated as relevant in defining himself and the others around him.

Antonio Balduino and the general public of Morro do Capa Negro are enveloped by the drumming sounds that come from Jubiaba's macumba, like warrior sounds, like sounds of liberation. The street vendors become the gatekeepers of Jubiaba's terreiro. Since the terreiro itself is linked to freedom and resistance, the image of the street vendors is inverted in *Jubiaba* as the novel that inserts them into Afro-Brazilian tradition. About this dynamic:

"Night was falling at the back of the houses and it was that calm, religious night in Bahia de Todos os Santos. From the house of the father of saint Jubiaba came the sounds of atabaque, agogo, chocalho, cabaga, the mysterious sounds of macumba that were lost in the twinkling of the stars, in the silent night of the city. At the door, black women sold acarajes and abaras...[...]And all the black people who wanted to make despachos paraded before the father of the saint. Some were prayed to with mastrugo branches. That's how the city would fill up the next morning with things that cluttered the streets and which passers-by were afraid to move away from. Rich people often came, doctors with rings, rich people with cars." (Amado, page 95)

The audience at Jubiaba's terreiro is made up of people from different ethnic groups and even different backgrounds. The white man who attends the terreiro in a particular passage of the book may be an ethnographer or an artist:"[...] he had come from far away just to watch the macumba of father Jubiaba" (Amado, 1935, page 101).

The fact that a man comes from far away to take part in the ceremony indicates the uniqueness of these practices in this place. His presence also shows the comprehensive nature of the terreiro's reach. The man takes part in the ceremony like everyone else in the audience. The women danced around the room and greeted Jubiaba, the highest authority in the terreiro. Then they turn and greet the audience, who animated the saint, "that is, the orixas who danced through the Feitas" (Amado, 1935, p. 97).

From these times of rascality and territorial disputes, a scout discovers Antonio Balduino's potential as a boxer. This brings us to the opening of the narrative. At the same time as this period of victorious fighting, Balduino suffers the greatest defeat of his heart: he discovers that the owner of his platonical love Lindinalva is getting married. In addition to his disappointment, Balduino loses his boxing match and his taste for his lumpesinato. We come to the fifth movement of his journey: as he sinks into emotional despair, he leaves in search of the "way to the sea" on Mestre Manuel's sloop. However, even though he is emotionally wounded, Balduino goes in search of another dream he once had: life on the quayside with the adventures of the sea that would bring him a new world:

"Large canoes lay motionless on the still water. The sloops, sails down, slept in the darkness. Even so, they gave the idea of departure, of journeys through the small ports of the Reconcavo with their great fairs. But now the sloops were asleep [...] They were going to fill up with vegetables, fruit, bricks and tiles. They would run all the markets. They would then return laden with fragrant pineapples[...] Antonio Balduino knows the history of all these sloops and canoes. Ever since he was a boy, he's liked to lie here on the sandy quay, his little face on the sandy pillow, his feet in the water. The water is warm and delicious at this time of night.

Balduino is sometimes fishing, silent, his face opening up in smiles when he hooks a fish. But in general, he only looks at the sea, the ships, the dead city behind him." (Amado, 1935, p. 108)

By sea, Balduino dreams of romanticized adventures, but what he recognizes is once again the writer's realistic proposal for this book. He arrives in the tobacco plantations of Bahia's interior, where he experiences hunger and economic oppression. After the wandering stages, this is a phase of seclusion and suffering. As he is dispossessed of his place of belonging and existential comfort, other questions emerge. The time he spent on the farm, albeit brief, receiving poverty for a menial job, is an important part of the formation that will lead him to identify with and integrate into the emerging proletariat that he will find back in Salvador:

"And the hands that stooped to the earth, big, calloused hands that picked the fragrant tobacco leaves. The hands lowered and raised themselves in a certain rhythm that was always the same. They looked like people who were crying. And that work gave them a pain in the back, a thin, prolonged pain that lasted well into the night, hurting. Zequinha would pass by watching the servant, giving orders, fighting. Heaps of tobacco leaves were piled up and, when evening came, the men's hands had earned ten dollars that they couldn't see, because they already owed their bosses unknown sums." (Amado, 1935, p.141/142)

The sixth moment of the plot is the one that best condenses the aspect of mobility imposed on the character's life and the consequent definition of who he is becoming. After rebelling against the life that harms the rural worker, he decides to move on before languishing once again:

"Take the railroad bed. When he reached Feira de Santana, he found a truck that would take him to Bahia. And he'll be happy because of the adventure he's had, the fight he's sustained, the siege he's broken through. [...] Jubiaba always said that brave men turn into stars... And the black Antonio Balduino lets out a laugh that silences the crickets and frightens the animals in their burrows." (Amado, 1935, p. 162)

This movement starts in a train carriage and ends up on a boat, back in Salvador. In this train carriage, he recognizes in the strangers he meets some of the situations of oppression that he will face in the future: the lack of employment, the precarious working conditions, the work specialties of the prevailing social classes, and especially the differences in opportunities by gender. These are brief elements, but they are responsible for a great deal of reflection on the part of the protagonist. Even so, the train journey is a causality of Balduino's destiny: again by chance he meets his former boxing manager and goes to live the itinerant circus life. Another facet of the character:

"There's a whisper in the crowd. Luigi goes out and enters with the black Antonio Balduino, who is wearing a tiger skin over his muscular body that is too small for him and restricts his movements. He crosses his arms over his chest and looks at the spectators with a look of defiance[...] They've set up a circus in Sao Felix. The circus is entertainment for poor people and Sao Felix is a city of workers. A man appeared to fight Antonio Balduino. He was a black man who had once been a sailor. The fight was widely announced. Luigi rubbed his hands together, satisfied with life, and was already listening to Antonio Balduino's sambas without getting angry. The clown ran the town, the men commented, the women laughed. On the opening night, the front of the circus was lit up, the orchestra came with the kids behind, black women selling mungunza at the door." (Amado, 1935, page 200)

Balduino arrives as a hero once again to restore the success of the Circus. However, the failure that the circus was experiencing also represented the future itself: the departure from what would have been the remnants of a past that was being abandoned by progress in society. The precariousness of this situation, expressed in the financial indigence of the character and the decay of the circus, gives this moment a sense of via crucis: with each new village the circus arrived smaller and with fewer components, until it was extinguished with the death of the owner Giuseppe. On the day of the circus' demise, it seemed that even nature was understanding the moment:

"But winter was hitting the river. Thick rains blurred the face of the waters. The river was rushing downstream, carrying logs it had uprooted from the plantations, the corpses of animals and even a door that the water had taken from a dwelling. The stone crowns had disappeared and the men no longer entered the river to look for fish for lunch. The river was treacherous and rumbled like a wild animal. Groups watched it from the top of the bridge and it passed underneath like a snake. From above came the sweet smell of smoke. The river had already swallowed two sloops this winter. There was a bereaved worker in one of the factories." (Amado, 1935, p. 206, 5p)

In these six movements of Balduino, Amado has so far chosen a repetition of situations: the conflict escalates, the momentary equilibrium is broken and then the heroic protagonist is reborn, grown up and reinvigorated in these stages that delineate the man Balduino has become.

The last movement will mark the meeting of Balduino and Lindinalva, a moment that is more transformative for the character than an amorous encounter. On discovering that the young woman who has always been his platonic love is prostituting herself out of necessity to raise a child, Balduino experiences a serious identity crisis: Lindinalva, on the verge of death, begs the black man to look after her son when she is gone. At this moment, when Lindinalva recognizes Balduino's virtues, the trickster dies in him and the adult is born, the new father and, later, class consciousness. He has to leave his job as a dockworker on the quayside to take care of his young son. And the vicissitudes of life show him the path of the class struggle. In the strike moments to come, Balduino feels born again and in the assemblies he won't make a speech, but "tell" the story of his search to be recognized as a leader. To raise the boy, Balduino becomes a dockworker on the docks of Salvador, which leads to his active participation in the proletarian movement. When the electric streetcar workers go on strike, other groups of workers from other sectors, including Balduino and his fellow dockworkers, support them. The general strike brings the whole city to a standstill. Balduino learns from his leaders how the proletarian movement is a string of beads: "if one of the pearls falls, the others fall too" (Amado, 1935, p. 290).

Within this symbolism of life and death, beginning and end, a new identity will emerge for the character who sees his destiny intersect with that of the child and, at the same time, with that of his class. Throughout this trajectory of movements that define who the protagonist Antonio Balduino becomes, some issues become common in all the phases: "all the Amadian character wants is 'not to be a slave', and this search for freedom leads him first to naughty rebellion and then to militant operaha" (Assis Duarte, 1996).

Another important analysis to be made about this latest movement by Antonio Balduino is that once again the setting is not merely an aesthetic component, but an important character who will also help sculpt him in this phase of his identity reconstruction. The sea, the dockworker's faithful companion and a long-time love of Antonio Balduino's, once again plays an important role in defining him:

"The huge shadows of the cranes appear on the sea. And the green, oily sea calls out to the black Antonio Balduino. The cranes make slaves, they kill men, they are the enemies of the blacks and the allies of the rich. The sea sets them free. It will only take one dive and you'll have time to let out a laugh. But Lindinalva stroked his head and asked him to look after her son [...]. Antonio Balduino had spent the night unloading a Swedish ship that was bringing material for the railroad and which would have been full of cocoa the following nights. [...] The cranes deposited huge rolls of iron on the quay. The blacks went under them to the warehouses, looking like strange monsters, and still talked." (Amado, 1935, p. 245/246)

Antonio Balduino, now a dockworker and a responsible father, is the one who, after many transformations, is chosen by Amado to be the individual who draws class consciousness from the strike. In the past, the feeling of revolt had already taken hold of Balduino, especially because of the discomfort of order and the possible absence of freedom. Today, the context has left the plane of idealization and found its daily life. First, the spatio-temporal conjuncture of before and now:

"They leave in groups and at the door Antonio Balduino remembers a man who was arrested there when he was giving a speech. He was a street kid but he remembered it perfectly. He had shouted, and the whole group with him, protesting against the man's arrest. He had shouted because he loved shouting, booing the police, throwing stones at soldiers. Today he needs to shout again, just like in the days when he ran free in the street and didn't see the enemy cranes ready to blow his head off." (Amado, 1935, p. 247)

Then came the birth of class consciousness:

"[...JAntonio Balduino had always had great contempt for those who worked. He would have preferred to go into the sea and commit suicide one night on the quay, rather than work, if Lindinalva hadn't asked him to look after her son. But now the black man had a new respect for the workers. They could stop being slaves. When they wanted to, no one could stop them.[...] But they laugh because now they know they are a force. Antonio Balduino discovered this too and it was like being born again." (Amado, 1935, p. 251)

In this last ontological element of the protagonist's definition, Balduino jumps from malandragem to militancy and sees the strike as the key to achieving a social identity free from the vestiges of slavery.

4 - Antonio Balduino and the city of Salvador:

Reading the pages of *Jubiaba*, in all of Balduino's movements through the places as well as in all of his temporalities - from childhood to adulthood - one element of this trajectory is repeatedly highlighted: the city. This urbanization that defines the city of Salvador is not merely a descriptive aspect of the landscape, but another geographical foundation of the figure of Antonio Balduino.

Rossi (2009) points out several times in his analysis of Jorge Amado's literature in the 1930s - where the book *Jubiaba* fits in - that it is important to realize how much the writer inaugurates regionalism from another angle: that of urbanization. The concept of regionalism in literature is much discussed, and often even defended as a way of looking more at the places and distances of the country portrayed than specifically at the literary work itself, where it would also represent the exotic (Candido, 1999).

Regionalism had a very emblematic theme: agrarian issues in the northeast with a strong association to the reading of rural misery. In Amado's works from the 1930s, this was not his bias, thus making him a constructor of the real possibility of the existence of an urban form of regionalism. The issues may have points of intersection and the same references, but the spatial and functional nexus was definitely not the rural but the urban.

In Amado's work, the city became an element that decisively marked his novels and in *Jubiaba* it is easy to see why the city is so important. Balduino, in almost all his moments of reflection on his being and who he was, does so under the aegis of observing the city of Salvador. Even though this occurs in his memories when he was far away. The Amadian city is the one that forms the privileged spaces for the characters to act and become aware of themselves:

"The city stretched its arms from the churches to the sky. From the quayside he could see the hillsides, the huge old houses. The lights shone above and white clouds ran through the sky like flocks of sheep... He lost himself in looking at the black houses of the city. There was a star right above his head. He didn't know what it was, but it was beautiful, big, shining in a blinker. He had never seen such a star before. The moon appeared very large and cast such a strange light over the back of the houses that he never saw the town again. He thought he was a sailor and had arrived in a foreign port. A faraway port like the ones he sees in his dreams every night. Because every night Antonio Balduino dreams that he lands in lands of other countries. Clouds were racing across the sky. They were rams. White, huge rams. There was no one in the Lower City. It was also the first time he had daydreamed like this. Bahia was no longer Bahia and he was no longer the black Antonio Balduino, Baldo, the boxer who used to go to Jubiaba's macumbas and who had been beaten up by Miguez, the Peruvian. What city would that be and who would he be? Where had all the people he knew gone? He looked towards the port and saw the ship. Of course, it was already time to get off, and they were waiting for him on board." (Amado, 1935, p. 109)

For Amado, it is the urban experience that enables the characters to understand the social world, its conflicts and contradictions, as if the rural world alone were not enough to have its "intellectuals", who, in order to be considered as such, need to move to the city. And those who have experience of the mass workers' movements or join the ranks of the "proletarian party" will draw more on the self-knowledge that only emerges from urban life.

In *Jubiaba,* when the city takes over the landscape and Balduino's thoughts, the Bahian writer proposes new sociological, political and aesthetic dilemmas. The Salvador depicted in Jubiaba contains the oppressive elements of the capitalist system and also alternative spaces, which are not utopian, but foreshadow the writer's communist city. They are alternative spaces because they resist the order of the city in one way or another.

According to Sellman (2013), the city is very permeated by these so-called spaces of resistance and in order to better understand the dynamics of how these places function, he problematized them using a Foucauldian concept of heterotopia :[14]

"real places - places that do not exist and that are formed at the very foundation of society - that are something like a counter-order, a kind of utopia effectively enacted in real places, all the other real sides that can be found within culture, are simultaneously represented, contested and inverted" (Foucault, 1966).

These places are located outside the basic systemic flow of the geographical environment of this urban space, although they are situated within the city; they represent part of the city and are something more. They offer a mirror to the real aspect, but they are different from utopia, which is a non-space: heterotopias are also real; they represent the otherness and sameness of modern places. The rest of the city must eventually recognize these spaces. They feed on the urban continuum, reversing, reshuffling, and experimenting with elements that have been taken from traditional places.

In *Jubiaba,* some elements in the landscape can easily be identified as heterotopia: the Morro do Capa Negro and, in a closer spatial cut, the terreiro de pai Jubiaba. These, in turn, are places of great ambience for Balduino, which are in the city, but are not the city itself according to the current functional order. Let's start with Morro do Capa Negro:

"[...]Don't you know why this hill is called Capa Negro? It's because this hill used to be this man's farm. And he was an evil man. He liked black men to have children with black women so that he could earn slaves. And when a black man didn't have a child, he had black men captured. He captured a lot of black people. Bad whites... That's why this hill belongs to Capa Negro and has a werewolf on it. The werewolf is a white man. He didn't die. He was too bad and one night he turned into a werewolf and went out into the world scaring people. Now he's looking for his home, which was here on the hill." (Amado, 1935, p. 36)

Morro do Capa Negro is an important point in Balduino's life. His first references to belonging were born there. The legend about the name of the Morro do Capa Negro, even though it comes from a past context, was well referenced when he said that the werewolf of the old legend was the white man of today. At this stage, Amado believed that social and racial struggles had exclusively economic causes. These issues become more apparent in Salvador, a city with the largest black community in Brazil and an economy in difficulty at a time when it exacerbated social inequality. Afro-Brazilian culture was also thriving in the city, despite repression from the authorities. These characteristics

Concept developed by Michel Foucault in a lecture to around 1900 architects (1967).

made Salvador an apt representation of the centuries-long conflicts in Brazil and the marginalization of a significant ethnic group. Thus, we see in the writer's choice of the myth of Morro do Capa Negro also a way of mentioning these issues.

The structure of the novel is very important for the construction of Salvador within it: due to its Manichean dynamic, *Jubiaba* presents a Salvador divided between oppressors and oppressed. From his position on top of the hill, Balduino gathers moments of Salvador's daily life to build the city in his imagination. For him, it is a place of 'life and struggle' (Amado, 1935, p. 16).

The sounds of suffering he hears reinforce this perception of how they mix with sounds of joy, voices and machines. Take a look at the following passage:

"The nights at Morro do Capa Negro were very pleasant. Antonio Balduino learned a lot in his childhood, and especially a lot of history. Stories that men and women told together in front of their neighbors' doors during long conversations on moonlit nights.[...] Life on the hill of Capa Negro was difficult and hard. All those men worked hard, some on the docks, loading and unloading ships, or carrying travelers' suitcases, others in distant factories and in poor trades: shoemaker, tailor, barber. Black women sold rice pudding, mungunza, sarapatel, acaraje in the city's twisting streets, black women washed clothes, black women were cooks in wealthy homes in posh neighborhoods. Many of the boys also worked. They were shoeshine boys, carried errands, sold newspapers. Some went to family homes and were raised in beautiful houses. The others would run around on the hillsides fighting, running and playing. They knew their fate from an early age: they would grow up and go to the docks where they would be bent over under the weight of sacks full of cocoa, or they would make a living in the huge factories. And they didn't rebel because it had been like this for many years: the boys from the pretty, tree-lined streets were going to be doctors, lawyers, engineers, merchants, rich men. And they were going to be the servants of these men." (Amado, 1935, p. 30/31)

The city emerges as a mythical place for Balduino where the hero will have to prove his strength against so many examples that surround him. Seen from the top of Morro do Capa Negro, it has a fearful but also seductive feel. The hill represents the high space of honest poverty and childlike purity of the hero protagonist. Below it, the city resembles the demonic, labyrinthine world that Baldwin will have to master in order to establish his virtue over the dangers that threaten him.

Some of the features of Salvador's cityscape portrayed in the book also help us to understand Antonio Balduino's life references even more, and what shapes the dynamics of his surroundings. Among the most distinctive features of Salvador's landscape is the city's dominant religion, Catholicism. This contrasts even more with the references to animist religious matrices, such as the Jubiaba terreiro, and further affirms its heterotopia. Its historic architecture reflects a colonial-style socio-economic system; and the large human group of oppressed people who sustain this system, such as the majority of the inhabitants of Morro do Capa Negro. Catholic churches in the city bear the marks of an exploitative order: gold and other precious stones that were extracted by black slaves in Brazilian mines during the 17th and 18th centuries. As such, these churches stand as traces in the landscape of colonial power with which the Catholic Church was associated. In this sense, the churches are extensions of the colonial houses, which also appear in the novel as the remains of an oppressive system: the 'Sobradoes' are the leftovers of the elite who had moved to noble areas of Salvador. The old colonial system has simply metamorphosed in the city, but it hasn't disappeared.

Even so, the description of the landscape contains many variations on the theme of the old legacy of colonization. This suggests that the city is stuck in its past. Modern slaves in the city no longer live in the slave quarters, but they have to survive in the old colonial houses on low wages.

Returning to Morro do Capa Negro, Balduino's childhood was spent there and then he left. When he returns to the Morro later and with a much changed self, it is clear that Balduino needs to look for his references:

"The city stretched its church spires into the sky. From the quayside he could see the hillsides, the huge old houses. The lights shone above and white clouds ran through the sky like flocks of sheep... He lost himself in looking at the black houses of the city. The moon appeared very large and cast such a strange light over the back of the houses that he never saw the city again. He thought he was a sailor and had arrived in a foreign port. A faraway port like the ones he sees in his dreams every night. Clouds raced across the sky. They were rams. Beautiful, huge rams. There was no one in the Lower Town. It was also the first time he had daydreamed like this. Bahia was no longer Bahia and he was no longer the black Antonio Balduino, Baldo, the boxer who used to go to Jubiaba's macumbas and who had been beaten up by Miguez, the Peruvian. What city would that be and who would he be? Where had all the people he knew gone? [...] The noise was growing on the hill. It came like a plea, like a cry of anguish. He saw, then, that the city was Bahia again, very Bahia, which he knew all [...] There were sounds of drumming coming down from all the hills, sounds that on the other side of the sea had been warlike sounds, drums that resounded to announce battles and shit. Today they were sounds of supplication, slave voices calling for help, legions of blacks with their hands stretched to the heavens. Some of those blacks who already had white faces bore whip marks on their backs. Today the macumbas and candombles send out those lost sounds. It was like a message to all blacks, blacks who were still fighting and shitting in Africa, or blacks who were groaning under the white man's whip. The sounds of drumming came from the slums. They were also anguished and confused, religious sounds, warrior sounds, slave sounds, to Antonio Balduino who was lying on the sand of the pier. The sounds entered his ears and rattled the deaf hatred that lived inside him." (Amado, 1935, p. 110)

The moonlight obliterates the electric lights that Balduino associates with Salvador. This new brightness and its accompanying tranquillity also distance him from the city. Balduino doesn't recognize the city and, by extension, he doesn't recognize himself either. Slowly, a familiar noise transforms his perception of the place: the sound of distant drums. These drums are played in candomble terreiros on hills such as Morro do Capa Negro, slums where black people make up the vast majority of the inhabitants. Jubiaba makes the terreiros as characteristic of the city as the "black houses of the urban space". However, the drums do not reinforce the colonial order that these houses represent. The sound of the drums makes Balduino recognize that he is an integral part of a proud group, the black inhabitants of Salvador, who have been victimized and enslaved in many different ways since colonial times. The colonial urban space of Salvador metaphorically stifles the sounds of war from the terreiros, just as white masters suppressed rebellions by black slaves. The aspect of the terreiros' resistance frames the proletarian struggle that could lead to a communist city in the book. The slum and the city are spaces for defining who the man Antonio Balduino is. If, at first, the slums provided the keys for Balduino to understand his oppression as a black man, then, secondly, his move to the city made him aware of the other shackles, apart from the racial ones, that he was supposed to carry in his life because of what he represents in Salvador and in his respective society.

This community experience in Morro do Capa Negro is replicated in other poor areas of

Salvador. There are other "poor people's parties" on the other side of town: street fairs and parties in Brotas, a rural area with both middle- and lower-class houses that is close to the central middle-class neighborhood of Nazare; Itapagipe, a suburban workplace and middle-class neighborhood; and Rio Vermelho, a more distant coastal district (Amado, 1935, p. 92/93).

According to the reading of the Amadian text, in any case, these locations reveal that the poor inhabit and use the marginal areas of the city more freely, while the central areas belong mainly to the upper classes. The rich live close to the city center, in airy districts like Nazare and Graga. Antonio, the owner of the Lanterna dos Afogados, a bar in the Lower City on the seafront frequented by the characters in the book, dreams of having a café in the center of the city. Thus, in addition to identifying these locations as heterotopias - appropriating Foucault's concept here - we glimpse in these areas frequented by Balduino(s) where the true roots of Salvador's popular resistance, so characterized by Amado, would be.

On the other hand, as a place where social classes mix, the Cidade Baixa is an important place for the poor, especially the Agua de Meninos market:

"The Agua dos Meninos Fair starts on Saturday night and lasts until midday on Sunday. The canoeists moor their canoes in Porto da Lenha, the sloop masters leave their boats in the small port, men arrive with loaded animals, black women come to sell porridge and rice pudding. Streetcars pass nearby, full of people. Everyone comes to the Agua dos Meninos Market." (Amado, 1935, p. 211)

The Agua de Meninos fair is a party. A black party, with music, guitars, laughter and fights. The stalls stretch out in rows. But most of the things are not on the stalls but in big baskets, in troughs, in crates. There's everything at the fair. *Agua de Meninos* simultaneously reinforces and breaks down the city's dualities. Firstly, it's more than a market. Originally a commercial area, it is actually used in several facets of the writer's narrative. People go there to trade, but also to talk and have fun. In addition, the novel relates this mixture of uses to the black citizens of Salvador and their practices: "Festa de Negro." The narrator summarizes the description of the market with the word "festa" to explain that this is a place for celebration while the name of the place indicates its commercial function (Sellman, 2013, p. 125).

Salvador is - or should be - a city where different cultures intersect. Jubiaba, the highest representative in the city's candomble hierarchy, also stands alongside the orixas as the city's patron or protector. The orishas bless the city, but Jubiaba, the "father of saints", supervises the orishas in their interaction with it. Jubiaba is a gatekeeper at the intersection of the human and mythical worlds, which is located in his terreiro on Morro do Capa Negro. Here, another element of resistance is configured and the figure of Balduino is formed, representing the social subject collectively.

Thus, when Balduino draws on the practices and links he learned in Morro do Capa Negro and especially in Jubiaba's terreiro, he succeeds where the strike leaders had not been before; the Manichean message they propose - oppressed employees against oppressive employers - is not enough, but Balduino's images become stronger symbols of workers' resistance and unity, once again indicating the work's Marxist bias.

In a segregated city, candomble terreiros have become revolutionary. They stand as what Edward Soja (1993) calls "spaces that make a difference", which are open and function outside the prevailing norm. As Soja explains, these spaces subject the original dualistic elements, such as raga and class, to a "creative process of restructuring" and open up "new alternatives". With these new combinations, such spaces have the potential to promote change in the urban continuum. Observe:

> "Night was falling at the back of the houses and it was that calm, religious night in Bahia de Todos os Santos. From the house of the father of saint Jubiaba came the sounds of atabaque, agogo, chocalho, cabaga, the mysterious sounds of macumba that were lost in the twinkling of the stars, in the silent night of the city. At the door, black women were selling acarajes and abaras...[...]And all the black people who wanted to make despatches paraded before the father of the saint. Some were prayed to with mastrugo branches. That's how the city would fill up the next morning with things that cluttered the streets and which passers-by were afraid to move away from. Rich people often came, doctors with rings, rich people with cars." (Amado, 1935, p. 94/95)

Soja's spaces of resistance existed, represented by Afro culture among other references. Even so, for Amado the ideal communist space is a city. *Jubiaba* highlights the role of Salvador as an avant-garde in the sense of a workers' revolution, especially when contrasting the city and its rural surroundings. In the Reconcavo area, Amado places Salvador in a privileged position: it is the right place for his communist ideal to begin. A future of hope, so to speak, only comes through the big city: "The Lights of Bahia are a Salvation" (Amado, 1935, p. 239).

5 - Jubiaba's macumba: the terreiro and Afro-descendancy

From Antonio Balduino's point of view, Jubiaba was one of the few free men in Morro do Capa Negro. Even so, he was persecuted for being a "macumbeiro". *Jubiaba*'s own terreiro is linked to freedom and resistance, and the novel also helps to insert him into Afro-Brazilian tradition. At this moment, the terreiro transcends the particular: it could be that of father Jubiaba, like any other that occupies the imagination and experience of the black population of Salvador.

There are many elements that demarcate the Afro-Brazilian roots in the Jubiaba terreiro. One of Jubiaba's macumba drums is an atabaque, for a tall man, and the Afro-Brazilian wooden hand drum that is also used in capoeira performances. Afro may be the Brazilian mix of dance and martial arts that Balduino learned as a child in Morro do Capa Negro. The atabaque symbolizes not only the cultural practices of the terreiro, but the culture and history that shape the identity of the entire Morro do Capa Negro neighbourhood.

By presenting the Afro-descendant elements present not only in the fictional Salvador, but also in the more real side of the city - this is the largest black community outside Africa - and where Jorge Amado gives greater vent to his dealings with black people. The form is almost always crude and primitive - at the same time introducing us to Afro elements that were previously exotic to the non-practicing masses - and in a rhythm interspersed with batuque, emphasizing popular orality:

"The audience was cramped around the room, close to the wall, their eyes fixed on the ogas who sat in a square in the middle of the room. All around the ogas were the made ones. The ogas are important because they are members of candomble, and the feitas are the priestesses, those who can receive the saint. Antonio Balduino was an oga.[...]The women's bare feet beat the clay floor, dangling. They ritually blew their bodies, but this blowing was sensual and sweet, like the hot body of a black woman, like the sweet music of a black man. Sweat was pouring down and everyone was taken by the music and the dancing. [...] Everyone in the room went crazy and danced to the sound of atabaques, agogos, rattles and cabagas. And the saints were also dancing to the sound of the old African music, all four of them dancing between the ogas." (Amado, 1935, p. 86/87)

The Candomble temple in Jubiaba has the function of being one of these marginal places in Amado's Salvador. This space rearranges Afro-descendant cultural elements in Salvador to present them in a potentially disturbing way. As presented, Candomble ceremonies invert everyday roles and hierarchies. The ordinary citizen has the chance to be the extraordinary. These places of resistance are possibilities for signaling differences within the limits of the predominant place where they are generated, i.e. Salvador. The women danced around the room and greeted Jubiaba, the highest authority in the terreiro. In this movement we see the birth of both the religious syncretism portrayed in the novel and the importance of the character Jubiaba, who now justifies the title of the book:

"The orixala was Xango, the god of lightning and thunder, and this time he had taken a feita, the little black girl came out of the chamber dressed in the clothes of the saint: a white dress and white beads spotted with red, carrying a little stick in her hand [...] Then the saint entered the midst of the feitas and dangou too. The saint was Xango, the god of lightning and thunder, and he was wearing white beads dotted with red on his white dress. He came and bowed to Jubiaba, who was among the ogas and was the greatest of all the pais de santo. On the Catholic altar, which was in a corner of the room, Oxossi was Saint George; Xango, Saint Jeronimo; Omolu, Saint Roque and Oxala, the Lord of Bonfim, who is the most miraculous of the saints of the black city of Bahia de Todos os Santos and of the father of saint Jubiaba. He has the most beautiful feast, because his feast is all like candomble or macumba." (Amado, 1935, p. 87)

It is also important to identify the religious traditions and their respective myths linked to Afro-descendant culture as elements in the formation of man. Why has this movement become so strong in the city of Salvador? Prandi (2001), an active researcher into the culture of the orixas, said:

"Since the 1960s, traditional religions have experienced a significant revival, among them the religions of the orixas established in America, with the great expansion of candomble, which spread from Bahia throughout Brazil, and Cuban santeria, now also cultivated in the United States, especially among Hispanic-American immigrants" (page 19).

The expansion of candomble in Brazil involved a strong adhesion of social segments different from those in which the religion of the orixas originated in the country, with the inclusion of adherents not necessarily of black origin and who come from social strata with more schooling and accustomed to the idea of information by book. I believe that, influenced by this cultural melting pot, Jorge Amado ended up coming into contact with these sources. In his book *Mitologia dos orixas* (*Mythology* of the Orishas), Prandi (2001) documents stories and histories of the culture of the orishas, identifying their myths and allowing an association with elements that Amado inserted into his narrative. Analyzing the role that the myths of the orixas leave as a legacy to Salvadoran culture in this research involves analyzing the figure of Jubiaba as the father of the saint who protected the inhabitants of Morro do

Capa Negro and, with his religious rites, unfolded the mythic present there.

Exu is a character who appears repeatedly in Amado's books. In *Jubiaba* he is even compared to Antonio Balduino himself. From a mythical perspective, Exu is a messenger. In the history of this messenger orisha, he had before him all the knowledge necessary to unravel the mysteries about the origin and governance of the world of men and nature, about the destinies of men, women and children and about the paths of each person in the daily struggle against the misfortunes that threaten each of us at every moment.

The way in which Amado narrates the elements of candomble rituals and the terreiro of the father of saint Jubiaba clearly shows this mythical bias. Prandi (2001) argues that:

"the myth is impregnated in the ritual objects, in the songs, in the colors and designs of the clothes and necklaces, in the secret initiation rituals, in the dangas and in the very architecture of the temples and, markedly, in the archetypes or models of behavior of the child of the saint, which recall in everyday life the characteristics and mythical adventures of the orixa from which the human child is believed to be descended" (Amado, 1935, page 19).

Each orixa can be worshipped according to different invocations, which in Brazil are called qualidades and in Cuba, caminhos.[...]. Each orixa multiplies into several, creating a diversity of devotions, each with a specific repertoire of rites, songs, dangas, vestments, colors, food preferences, the meaning of which can be found in the myths. In *Jubiaba,* Exu is the orixa who is always present, because the cult of each of the other orixas depends on his role as messenger. Without him, orishas and humans cannot communicate, and it is sometimes the father of saint Jubiaba and sometimes Antonio Balduino himself who carries out this mission.

In Brazil, where the oracular institution based on the figure of the babalao has disappeared, certainly because of the centralizing role played here by the maes and pais de santo, heads of the terreiros that bring together the devotees of the orixas, the myths have remained diffuse in the ritual memory and day-to-day life of the Yoruba descendant religious congregations. In traditional Yoruba society, a non-historical society, "it is through myth that the past is reached and the origin of everything explained, and through myth that the present is interpreted and the future predicted, in this life and the next" (Prandi, 2001, p. 24).

From the mid-1930s onwards, writers and social scientists began to record orisha myths more systematically, although a few examples date from the turn of the 19th to the 20th century. In the 1930s, the anthropologist Artur Ramos believed that Yoruba mythology in Brazil had been completely degraded and lost (Ramos, 1935), but Roger Bastide (1945, 1958), a French sociologist who was then professor of sociology at the University of São Paulo, researching in Bahia in the 40s and 50s, perfectly discerned the living presence of myths, not only as a narrative, but as a substratum in the rites held in the terreiros, especially in the dangas, and in the very mental structure of the followers of the religion of the orixas, having recorded countless rites. For Prandi (2001):

"In candomble, xango and other regional variants of the religion of the orixas in Brazil, as well as in the religion of the orixas in Africa and Cuba, myths justify the roles and attributes of the orixas, explain the occurrence of everyday events and legitimize ritual practices, from initiatory, oracular and sacrificial formulas to the choreography of sacred dances, defining colors, objects, etc. The association with any of these aspects is what gives life to the myth, and its proof of meaning." (page 32)

According to the ideas of Prandi (2001), the Yoruba believe that men and women are descended from the orishas, and therefore do not have a single, common origin, as in Christianity. Each one inherits from the orixa from which they come their marks and characteristics, propensities and desires, all as told in the myths. The orishas live in their domains, using all sorts of tricks and wiles, from covert intrigue to open and bloody warfare, from love conquest to treachery. The orishas rejoice and suffer, win and lose, conquer and are conquered, love and hate. Humans are just faded copies of the orixas from which they descend. Prandi's perspective allows us to build an even more ontological view. Since men and women are descended from the orixas and the orixas exhibit the same behaviors and sensations, myth and the human here merge into the same definition.

In *Jubiaba,* the writer's contradiction lies in offering a discourse on black religious rites as a form of alienation presented at the end of the novel in opposition to the class consciousness of the strike, which ends up being at odds with everything he describes in previous pages. Even considering this contradiction, other points of reflection can be made from this problematization of Afro-descendant culture in Salvador. In *Jubiaba,* the question of blackness comes up every time we think about the role of Antonio Balduino and his fellow dockworkers or other workers, since it is not simply a question of talking about the proletarian, but about the black proletarian. The relationship between Antonio Balduino and Jubiaba is permeated by several moments of these reflections throughout the narrative. And, from time to time, the roles alternate. And class consciousness is born in the black Antonio Balduino:

"Jubiaba doesn't age. How old will he be? He must be over a hundred. He also knows so much. Jubiaba increases the anguish that takes Antonio Balduino from time to time. Jubiaba says things that stay inside the black man and make him think of the sea where Viriato threw himself in, where old Salustiano forgot his children's hunger. Antonio Balduino thinks that he's not the same, that he's not as happy as he used to be.[...]It's the eve of Sao Joao and everyone is happy in the streets. Antonio Balduino would like to be cheerful too. The men passed him with smiling faces and the fire houses were full of customers. Everyone was preparing for a happy night. They'd be releasing firecrackers and starlets. The blacks only talked about Joao Francisco's party and the Liberdade ball in Bahia. However, Antonio Balduino couldn't be happy tonight. Clarimundo is dead and all he can think about is the stevedore." (Amado, 1935, p. 223)

Although the Amadian text represents Umbanda as a form of cultural resistance by blacks and even denounces the religious non-acceptance of which they are victims, it ends up framing blackness within the partisan discourse, whereby economic determination makes individuals equal, regardless of creed or color. It is at this intersection of Antonio Balduino's elements of class consciousness and the question of black race with Afro-descendant elements that the theme of Marxism emerges in the work mentioned.

6 - Amado's Marxist noopo in Jubiaba:

Jorge Amado's Marxism is a religious Marxism, as much or more than a political Marxism, as contradictory as that may seem. *Jubiaba,* through the experience of the strike, transforms the protagonist ontologically in yet another of his changes: from being seen as black until then, he is now seen as a worker demanding improvements for his social class. And this transformation has a spatio-temporal milestone: the speech that Balduino gives inside Jubiaba's yard on the first day of the strike:

"My people, you don't know anything... You need to see the strike, go on strike. Black people go on strike, they're no longer slaves. What's the point of black people praying, black people singing to Oxossi?[...] What can black people do? Black people can't do anything, they can't even dance for saints. Blacks go on strike, blacks for everything, for cranes, for streetcars, where's the light? There's only the stars. Black is light, and the streetcars. Blacks and poor whites are slaves, but they have everything in their hands. You just don't want to, you're no longer a slave. My people, let's go on strike because the strike is a necklace. Everything together is really beautiful." (Amado, 1935, p. 255/256)

It is in the strike that Balduino glimpses the possibility of re-signifying the mythical and legendary of Afro-Brazilian culture, "politicizing" it and thus incorporating it into the universe of struggle and class consciousness. Furthermore, the metamorphosis of the meaning of struggle as a very important element in Balduino's ontological construction is once again reinforced here. Boxing matches are left behind in this context. We now move on to the class struggle, a rich theme in Amadian literature at the time.

The powerful racial issue in *Jubiaba* is explained throughout the novel in Marxist terms: it is a class issue. By reducing the racial question to material terms, Amado presented communism as an umbrella solution for all social shortcomings in Brazil.

Greatly influenced by his aforementioned historical-literary context of socialist realism, Jorge Amado was interested in making a *mass novel.* The expression "*mass novel*", which Amado uses more than once, does not refer to what we now call *mass culture,* a later phenomenon. But "it had to do with the Marxist notion of mass, and with the desire to discuss social problems to which the same class was subject in different national contexts." (Machado, 2006).

Another important analysis is that in *Jubiaba* the capitalists are the villains and the masses are the heroes who must reclaim their degraded city from the oppressive elites. Using these ideas, Amado highlights the contrast between the capitalist order in the city and the promise of a bright future of socialist order. Popular stories were Jorge Amado's true school. Bahia is at once one and the other, misery and poetry, where dualities - such as oppressed and oppressors - also help to construct the Amadoan city.

In one of the last chapters of Amado's book, which is being analyzed for the moment, the narrator never refers to Salvador by name: it becomes just another city, among many in a network for an international proletarian movement. All the Manichean elements of the city must be replaced by a "one-party communist" order. However, this Manichaeism is a defining point throughout *Jubiaba*'s trajectory. Manichaeism (white-black; upper-lower; poor-rich; rise-decline) governs every minute of

Balduino's life.

Another important aspect of this Marxist legacy to be reflected on in *Jubiaba is* the celebration of freedom and the love of marginalized characters. Machado (2006) even raises the possibility of seeing anarchist aspects in the same text:

"It is in the search for personal freedom that this tendency manifests itself, above all with characters who break with the parameters of the straight life that society offers them and choose to follow their own paths, inventing original patterns that almost always shock and scandalize the well-thinking. And they do this with the instinctive naturalness of those who just want to be happy and can't conceive of an existence that isn't primarily faithful to freedom, without conventional brakes imposed by social roles."

In Antonio Balduino's life, this anarchist side appears in his lumpesinato movement. However, it is a movement with finitude. Naughty freedom ended up transforming Balduino's being into workers' militancy. "Balduino jumps from malandragem to militancy and sees the strike as the bridge to the conquest of a social identity free from the remnants of slavery" (Assis Duarte, 1996).

By the end of the book, Antonio Balduino will be classified as a hero of the emerging proletariat, a human figure who is a synthesis of his people, and the scene of the strike in this context is the main defining element of the protagonist's new being:

"The strike had once again been a real revelation for the black Antonio Balduino. At first he had loved it as a fight, as noise and brawling, things he had enjoyed since he was a child. Gradually, however, the strike began to take on a new aspect for the former boxer. It was something more than noise and fighting. It was a fight directed towards an end, knowing what it wanted, a beautiful fight. There, in the strike, everyone loved each other, defended each other and fought against slavery." (Amado, 1935, p. 281)

This scene invites two alternative readings. The first reading follows Amado's application of the aforementioned Marxism to the novel: after his experience with the strikers, Balduino is fully educated. He is now aware of the class struggle in Salvador. He knows that racial problems in Brazil stem from economic and social exploitation. And he finally understands how he should interpret the moral of the ABC stories - popular stories about brave men, which are usually told in verse, from the name "abc", in reference to their rhyming system - he had heard all his life: the proletarian movement must be the "right way".

By combining Balduino's final journey with the birth of a combative class, which inaugurated a new stage in relations between capital and labor in Amado's Bahia, *Jubiaba* comes close to this model. Antonio Balduino completes the definition of his being, becoming the representation of the man who emerges with the increasing inclusion of workers on the Brazilian political scene.

7 - The geography of Jubiaba:

As in the previous chapters, this final one will also focus on the geography of the book analyzed throughout this writing. However, for *Jubiaba*'s narrative, we will approach it from a theoretical perspective not yet mentioned throughout the text. This new theoretical argument has the same semantics as the geographic epistemology already discussed, but the focus will be different: the relationship between man and technology.

Another clarification about the conclusion of this chapter is pertinent: why is there a greater focus on technology specifically in the book *Jubiaba?* Because in the empirical understanding of this research, this would be the book in which we best perceive the ontological formation of man surrounded by the question of the technification of the city. As has already been argued, Antonio Balduino is a being whose formation process is accompanied by movement through the different environments he passes through. And along this path, the city of Salvador, at the time of the novel's spatio-temporal portrayal, represented a city being transformed by technology. And by a society driven in its construction by a technical environment.

Here we will use the theoretical propositions of Friedmann (1968) in his book "7 studies on man and technology", and his first argument about these ideas is present in the following passages:

"Man is not the same, he does not feel, act or think in the same way according to the epochs of his history, according to the environment in which he lives, according to the techniques at his disposal." (page 15).
"The mentality of the individuals in a human group is inseparable from the totality of their conditions of existence and particularly from the state of knowledge of the techniques and the language they have at their disposal to express themselves." (page 25)

Friedmann's (1968) first reasoning stems from the defense that we will understand the definition of man's being through his epochs of experience, which correspond concomitantly to a specific period of technology. In other words, the multiform Balduino, who could perfectly well be any man portrayed in the period in which the writer produced his work, is a being who defines himself according to the logic of technical apprehension that he develops with the environment in which he lives(u). Technical knowledge is a conditional factor for his existence as it unfolded. Another point about this Friedmannian perspective of theoretical reasoning is that technology is not exactly born in the technical environment, it is already a conditioner of the previous natural environment itself, where it actually inaugurates the expression of its existence with the man who works. See:

"[...] when we use the term 'natural environment', we do not forget that it has been a relatively technical environment since the origins of prehistory: *homo faber...* Man's efforts to defend himself, to feed himself, to shelter himself, to clothe himself, to move around, imply a progressive technical development for which, starting with the most crude societies, ethnology is today lending increasing interest" (Friedmann, 1968, p. 33).

It is important to briefly analyze the concept of the *natural environment,* since it is the starting point for the concept of the technical environment that we are working on in this chapter. For Friedmann,

"We call the natural environment, the environment of civilizations or pre-machinist communities in which man reacts to stimuli coming, for the most part, from natural elements, earth, water, plants, seasons, or from living beings, animals or men. In this environment, the various tools are direct extensions of the body, adapted to the body, shaped by the body according to processes in which biological, psychological and social conditioning are closely linked." (1968, page 76).

In *Jubiaba,* the effects of the transition from the natural environment to the technical environment are quite latent, such as the intense growth of the city of Salvador, which is treated indirectly but is perfectly perceptible in its effect through the dynamics of the characters. However, this technical environment portrayed in the book has an existential meaning that is different from the molds already established and worked on in geographical theories: it is the one in which man defines himself, being ontologically guided by the construction of his existence in it.[15]

For Friedmann (1968), the technical environment is that which develops in industrialized societies and communities. In this technical environment, the part of the stimuli that we defined earlier decreases and, at the same time, a network of complex techniques tending towards automatism is increasingly tightening around man. With this in mind, it's interesting to see Friedmann's (1968) concept of man formed from this logic of the technical environment:

"Today, the view of the world held by adults in cities cannot be explained solely by the general action of society, by 'collective representations'. It is essential to place them in each case in relation to the concrete place that the individual occupies in society and the particular characteristics of the technical environment." (page 47/48)

When we looked at the passages about Balduino, we could see that the place occupied by the protagonist was a defining element of the different facets or even of the now mentioned collective representations. However, it was at the moment when he experienced the phase of the striking dockworker that this technical environment most influenced his ontological definition.

In *Jubiaba*, every place mentioned - the party, the ball, the sea - fits into the situations of the collective representations. However, one in particular expresses well the idea defended here of man conditioned to the technical environment: *the cranes make slaves.*

Another passage that is even more explicit about the conditioning of man (Balduino) in relation to the technical environment is at the moment in the narrative when the protagonist gets involved in the strike movement. The technical environment is relevant in the contemporary reality of cities. In short, it seems to us that in order to explain the contemporary reality of cities and the countryside since the technicist period, it is necessary to bring into play - alongside socio-economic factors, of course, but in the foreground - the notion of the technical environment and therefore of psycho-sociological conditioning. Regarding the proposition that the subjects involved in strikes are inserted in

The concept of the technical environment has already been worked on by other authors in Geography, such as Milton Santos (1996). According to this author, this phase of the technical environment is characterized by pre-machine and machine techniques, corresponding to the post-industrialization period. In addition, this is the beginning of inland urbanization and the formation of a Concentrated Region. However, in this approach by Milton Santos, man was not ontologically formed by this technical environment as seen in Friedmann (1968). This is why this second author was chosen for this chapter's geographical problematization.

this context of the technical environment, another passage from Friedmann (1968):

"The instinctive theory is exclusively psychological and even psychoanalytical: in reality, the worker's attitudes at work depend on more varied and complex factors; they cannot be understood, as we have seen, except by analyzing the technical conditions of his job, which depend on the science of machines, by psychotechnics, which takes into account individual aptitudes, and by sociology, which interprets the worker's reactions according to his relationships with the various social and professional groups to which he belongs. It's not automatisms (real or supposed) that intervene alone in reactions to new working conditions, but the whole of mental life: it's the worker, and the whole man who is at stake, coming to terms with the new environment." (page 45)

Another interesting concept for geographical interventions based on the problematizations in Friedmann's book (1968) comes from the analysis of the genre of life. At the beginning of the 20th century, the father of French geography, Vidal de La Blache (1845-1918), was already developing this concept with a geographical slant. This concept will be very pertinent to geography in *Jubiaba.* Friedmann mentions this in his text:

"[...] the notion of genre of life is only a 'naturalistic' notion and, to a large extent, deterministic. Nature commands the behavior, the productive activity of a human group; however, the founder of the French school of anthropogeography, Vidal de La Blache, progressively insisted on the choice made by man within the resources he exploits or does not know about. He insisted on the techniques available to man and which are his work. In this way, Vidal freed the 'genre of life' from the strict determinations of natural causality." (1968, p. 78)

It can be said that the genre of life is perfectly endowed with a conceptual notion that can be integrated into both the natural environment and the technical environment as we understand them. The genre of life implies not only "the activities necessary for material subsistence", but also, "around the particular techniques that are gradually elaborated, a set of mental harmonies, customs, rites and social relationships" (Friedmann, 1968).

In *Jubiaba*, all of this is expressed in the narrative. It's the customs of Afro-descendant origin, it's the social relations between the workers in the class struggle, it's the rites of passage that Balduino experienced throughout his career inside and outside Morro do Capa Negro. It's the daily life of the people who live in Morro do Capa Negro, reinventing themselves there even though they repeat themselves every day. An infinite set of possibilities that bring their generos to life. The most interesting thing about Friedmann's (1968) concept of genre of life - in relation to La Blache's - is precisely its ontological position:

"It is possible to say that, in relation to the natural, relatively technical environment, the environment of contemporary industrialized societies comprises a number of technical elements of all kinds so immensely increased that the quantity of their effects is transformed into a new quality, and it is precisely this new quality that I propose to call the new genre of psycho-sociological conditioning of man by his environment." (1968, p. 80)

Based on Friedmann's ideas, let's go back to the conceptual proposition at the beginning of the thesis: the geographical foundation of man is also guided by the environment. Balduino, the protagonist of *Jubiaba, is* the ultimate living expression of this possibility of an ontological reading of man in a geographical context. An interesting passage of Balduino's that fits in with this "psycho-sociological" conditioning that we are also calling the "geographical foundation of being" occurs explicitly in his passage past the white man's house in Travessa do Zumbi:

"Silence and stillness came down from everything and went up from everything. They came from the distant sea, from the hills behind the unlit houses, from the lights of even the few streetlamps, from the people, they descended from the air above us and enveloped the street and its creatures. It seemed that night came earlier to Travessa Zumbi dos Palmares than to the rest of the city [...] The street was sad. An agonizing street. The calm of the street was weighed down by an air of agony. Everything around was in agony: the houses, the hill, the lights. The silence was harsh and painful. Travessa Zumbi dos Palmares was in agony." (Amado, 1935, page 42)

In fact, it was Balduino who was dying before anyone else. A kid from the slums who had been sent to be tied up in the Comendador's house. He was the one in agony, losing sight of his freedom. What's more, he associated it with that place. The house on the lane was the definition of the calm that weighed on his soul.

Another important consideration to be made about the application of the concept of genre of life is that the set of possibilities it can encompass in the case of Balduino necessarily involves his life in the city. In the space-time of the book, cities are agglomerations where certain types of life now predominate, giving this notion of the genre of life precisely this substance, which can be said to be elaborated, socialized. This notion of the genre of life is subject to the circumstances of the technical environment, to the conditioning of individuals by the technical environment, and also to "the re-appointments of individuals to this environment, the transformations of their sensibility and mentality in the new environment". (Friedmann, 1968).

Thus, the category of *urban lifestyle was* created by Friedmann:

"[...] It's an urban way of life: it's made up of men who are diversely conditioned, but all conditioned by the technical environment. I say 'diversely conditioned' because today certain studies in American sociology suggest the existence of types and tend towards a typology of conditioning by the technical environment." (1968, page 84)

Today's cities, on the other hand, no longer have strict boundaries, they are zones; on the other hand, they are conditioned by an increasingly clear, definable and scientifically apprehensible way of life. This was already the Salvador de Balduino presented by Amado, and which so defined his main character. Because of this relationship mediated by the genre of life between man (Balduino) and the city (Salvador), configuring in its formations a technical environment, we can also speak of a technicist civilization.

Technicist civilization gives rise to liberated time, clearly separated, at least apparently, from working time. This separation is commanded by the organization of work and its discipline, by the division of tasks, by the structure of factories, by the cohesion of the industrial armies that populate it. We close this chapter with the following proposal for reflection: Is Balduino part of this technicist civilization through his social role as a striker/worker in search of his liberated time or through his resistance and defence of freedom as the rascal from Morro do Capa Negro? I believe that the geography of this rich character is expressed in both situations.

Chapter 4 - Death and the death of Quincas Berro D'agua: a reading of man's being in the city of Salvador

In the mid-1950s - after two decades of commitment to socialist activism - events in the Soviet world led by Stalin made Jorge Amado re-evaluate his devotion to the political system of the Soviet Union. The denunciations of atrocities in the Soviet world made the writer put an end to his activist commitment. In 1953, with the publication of one of his most famous works, *Gabriela, Clove and Cinnamon*, he set a new course for his career, which began to explore sensuality and humor more strongly. The novel *A Morte e a Morte de Quincas Berro d'Agua* (1959) is part of this line.

The 1950s were also the time of Juscelino Kubitschek's developmentalism, a period of particular cultural effervescence. The prevailing optimism in Brazilian social and political life after the end of the Vargas dictatorship in 1945 may have contributed to the creation of the fantastical and comical atmosphere in the narrative of *Death and the Death of Quincas Berro D'Agua.* At times, the efforts of the writers of the 1930s generation to portray reality critically ended up restricting the scope for fantasy and imagination. Without giving up the sociological perspective, Jorge Amado reserved space for the absurd in some of his narratives. This is what happens in *Death and the Death of Quincas Berro d'Agua.* Like a good storyteller, the narrator develops his surprising plot, worrying less about being true and more about being verisimilar.

Analyzing this novel to close the final chapter of the thesis was not a random choice. After all the problematization throughout the previous three chapters and the theoretical clarification that accompanied them, the choice of *The Death and Dying of Quincas Berro D'agua is* because it is believed that there is no book by Jorge Amado with a greater nexus of ontological proposal than this one. Quincas Berro D'agua - here unquestionably mentioned in the book's title because the entire narrative revolves around this character - dies twice (or three?) in the plot. The first death is to inaugurate his redefinition as a being that will be characterized from then on in the text. And, once again, reiterating the construction of the social being through an urban space that perfectly aggregates our proposed symbiosis between man and the geographical environment. The city of Salvador is the one that constructs the geography of Quincas Berro D'agua.

It is possible to insert *Death and the Death of Quincas Berro d'Agua*, characterized by the contact between the world of the dead and that of the living. In fact, these two dimensions share the mystical space of the city of Salvador. Two sides of society are represented. On the one hand, the universe of the established order, with the proper framing of individuals in respectable social institutes, such as family, marriage and work. On the other hand, the universe of disorder, to which Quincas surrenders by rejecting the perverse logic that surrounded him, according to which marriage, family and work should be sustained even if they lead to unhappiness. Quincas rejects this logic to the end and even beyond.

Amado's style began to show an ironic detachment that emphasized "the relativity of truth".

This new approach, he argues, allowed Amado to question the dogmatism of both political and religious ideologies. This period marked a return to Salvador in Amado's fiction and his construction changed enormously: they portray Salvador as an inversion of his alienated city of the previous period. Some of the areas of Salvador that feature prominently are: the Pelourinho of Quincas and Itapagibe, still owned by the exemplary civil servant and family man of yesteryear, Joaquim Soares da Cunha.

The Death and Dying of Quincas Berro D'Agua is a dynamic book in terms of its plot and begins after Quincas' first death. This - a more moral than physical death - is the starting point for the subsequent characterization of Quincas' ontological redefinition. Therefore, in this chapter we will first analyze who this social being who (re)constructs multiple geographies known as Quincas Berro D'Agua is. Regarding this proposal for an ontological look at the work of *Quincas Berro D'Agua*, Fabio Lucas (1972) states:

"The social theme is compressed, bringing more individual subjectivity to the protagonist and less emphasis on the collective aspect of the characters. Everything revolves around a limit situation: death. This is the existentialist tone that emerges explicitly in Amado's work. Nobody knows whether Quincas Berro D'agua died once or twice. If Quincas died twice, he conquered death. Quincas would then be absolute."

In an ontological reading, agreeing with Fabio Lucas, it can be said that Jorge Amado then does metaphysics, circulates through phenomenology, which inquires into the essence of being. The novelist becomes more concerned with the intensity of the human emotions that form this being and less with their quantity. And, corroborating this reasoning, Vinicius de Moraes states in the book's preface that "if it is true to say that style makes the man, we have that Machado is more style than man and Amado is more man than style" (Amado, 1959, p. 11).

As identity cannot be constructed outside the social world, in this work, the author imprints spatial identity on the way his characters are and the way they exist. Quincas Berro D'Agua is the main character who defines the being of the man in question. We'll analyze this point below.

1 - O Quincas Berro D'Agua's "being":

Let's start with the title of the book: *The Death and Dying of Quincas Berro D'Agua,* a symbolism that in itself semantically allows us to create a dialog with the ontological reading of the character. Joaquim Soares da Cunha is a retired civil servant who leaves his comfortable middle-class home and his domineering wife and daughter to live among drunks, scoundrels and prostitutes in the streets of the old city, the Pelourinho neighborhood and its surroundings. This is his first symbolic death. From that day on, he became Quincas Berro Dagua, a legendary figure among the malandros of Salvador.

The circumstances surrounding the death of Joaquim Soares da Cunha, a civil servant, are surrounded by mysteries and disagreements. His family claim that he died a decent death, but hide some of the vexations of the last moments of his life. His friends, on the other hand, swear that Joaquim died at sea, as he had wanted. Take a look at the passage in the book about his first death:

"When a man dies, he is restored to his most authentic respectability, even though he committed follies in his life. With its hand of absence, death erases the stains of the past and the memory of the dead man shines like a diamond." (Amado, 1959, p. 18)

For most of his life, Joaquim/Quincas was an exemplary civil servant, with the respect of his colleagues and family. At the age of fifty, however, for an unknown reason, he said goodbye to his family with offensive words and started living on the streets. He spent ten years constantly drinking in the company of Salvador's scoundrels and prostitutes. His wife Otacilia couldn't resist the family drama and died. Daughter Vanda and her husband, Leonardo, had to put up with the existence of that troublesome relative.

On one occasion, the owner of the bar he frequented, wanting to play a trick on him, filled a glass with water and offered it to him instead of the usual cachaga.

Joaquim spilled the liquid and, upon realizing the deception, let out a screech that gave rise to his nickname: Quinoas Berro d'Agua. A name that will be defined by an experimental episode in which the continuity of the reference makes his being and daily life increasingly drawn in its place, redesigning the character's being. Take a look:

"[...] it was on that distant day that the nickname 'berro d'agua' was definitively incorporated into Quincas' name. He had entered the shop of Lopez, a friendly Spaniard, outside the market. A regular customer, he had earned the right to serve himself without the help of a waiter. On the counter he saw a bottle overflowing with clear, transparent, perfect cachaga. He filled a glass, spat to clear his mouth, turned it over and over. And an inhuman scream cut through the placidity of the morning at the Mercado, shaking the Lacerda Elevator itself to its very foundations. The cry of an animal wounded to death, of a man betrayed and disgraced: "Aguuuuuuuuuuuua!" (Amado, 1959, p. 58/59)

Quincas represents the contradictions of man and society portrayed by Jorge Amado. It is his break with the family and also a break with society. By breaking with the family structure, he denies his own social structure, that is, he breaks with the bourgeois structure. Does the fact that he was a 'good man' and then chose to be a 'scoundrel' deprive him of the status of a citizen? In this sense, a parallel can be drawn between Joaquim and Quincas: the former, an exemplary civil servant, husband and father, who showed a lack of "enthusiasm сото if it bored him and he didn't have the courage to say it" (Amado, 1959, p. 48).

As for Quincas, "we recall facts, details and phrases that can give him his just measure. He was the one who had looked after Benedita's three-month-old son for more than twenty days..." (Amado, 1959, p. 60).

The biggest change in Quincas is that disorder becomes an ideal of egalitarianism to be embraced. Quincas is always celebrating life in order to conceal his real precarious conditions. Take a look at the following passage of people reacting to his death:

"In the bars, in the botequins, at the sales counters and warehouses, wherever cachaga was drunk, sadness reigned and consumption was on account of the irremediable loss. Who knew how to drink better than him, never completely altered, all the more lucid and brilliant the more brandy he drank?" (Amado, 1959, p. 58)

When Vanda, his daughter, looks at Quincas' body in the box:

"He saw a smile. The cynical, immoral smile of someone who was enjoying himself. The smile hadn't changed, the experts at the funeral parlor had achieved nothing against it.[...] That smile of Quincas Berro D'agua had continued and, in the face of that smile of mould and enjoyment, what good were new shoes,[...] what good were black clothes, a clean shirt, a clean shave, starched hair, hands folded in prayer? Because Quincas laughed at it all, a laugh that grew wider and wider, that gradually resounded in the filthy pigsty. He laughed with his lips and his eyes." (Amado, 1959, page 50)

His second death, which the middle-class characters consider to be the official one, occurs in his sleep, in his impoverished apartment on Ladeira do Tabuao. His friends were not satisfied with this extremely bourgeois death of Quincas Berro D'agua. So they steal the body - or take Quincas depending on which version the reader chooses to believe - from the box on Ladeira do Tabuao to celebrate what they believe to be his birthday. After a night of merriment, heavy drinking and a bar fight, they sail off to open water. A sudden storm takes them by surprise and Quincas falls - or jumps? - into the sea and to his third and final death. (Amado, 1959, p. 173)

Quincas Berro D'Agua dived into the sea of Bahia and traveled forever, never to return. One of the biggest changes portrayed in Quincas' life, even though it was analyzed after his "deaths", is that the social "disorder" in which he lived re-founded his being. Thus, a realistic picture of Quincas' death would not be an obscure funeral, but a tribute at a bar table. Take a look at this passage in *Quincas:*

"The news of Quincas Berro D'agua's unexpected death was already circulating through the streets of Bahia. It's true that the small market traders didn't close their doors in mourning. On the other hand, they immediately increased the nails on the balangandas, straw bags and clay sculptures they sold to tourists, thus paying homage to the dead man." (Amado, 1959, p. 53)

As Quincas embodies human misery itself, stratified citizenship is evident. From this perspective, the author presents Quincas as a man who, at a certain point in his life, sets out to build a new existence. At the first moment of his life, Quincas was a man who conformed to standards, with no manifestation of his own will. In the second moment, he demonstrates to the world not only his desire to build his own identity, but does so by exercising citizenship in solidarity with his friends. This duality regarding how to view Quincas' death(s) is portrayed even at the beginning of the book:

"To this day, there is a certain amount of confusion surrounding the death of Quincas Berro D'agua. Unexplained doubts, absurd details, contradictions in the testimony of witnesses, various gaps. There is no clarity about the time, place or final sentence. The family, supported by neighbors and acquaintances, remain intransigent in their version of the quiet morning death, without witnesses, without apparatus, without sentence, which happened almost twenty hours before that other rumored and commented death in the throes of the night, when the moon broke over the sea and mysteries happened on the edge of the Bahia pier." (Amado, 1959, p. 15)

The conflict between the two versions of Quincas' death also becomes a dualistic conflict for other significant issues in the book's plot. The first of these, addressed by Sellmann (2015), is the conflict between the two opposing urban areas in the narrative: "homogeneous, orderly middle-class Itapagipe and heterogeneous, chaotic o Pelourinho lumpen." (page 185).

However, it is necessary to point out that the transcription of Quincas' spatial trajectory is situated in a very limited area of Salvador's urban area, which makes the representation of the city's binary "order versus disorder" too narrow to interpret the entire city. Therefore, we will now analyze how this ontological perspective of Quincas also makes us perceive aspects of the experience and definition of the city itself.

2 - Quincas and the city of Salvador:

Salvador is the city of Quincas Berro D'Agua. Salvador is the city that will define itself according to Quincas' multiple existences throughout the narrative. The landscape of Salvador functions as a dialectical setting for Quincas' life: the dynamics of the existential transformations of Quincas' being are portrayed and (re)constructed in the city.

Although extreme poverty remained in the city center, the rest of Salvador was drastically transformed at the end of 1950. Salvador was finally experiencing the effects of accelerated economic growth, which began with the discovery of new oil reserves in the Reconcavo Baiano area of the metropolitan region. At the beginning of 1960, the federal government began an ambitious investment program in the Northeast region of Brazil, through Sudene (Superintendency for the Development of the Northeast). The state of Bahia received most of this investment, which translated into many new jobs in Salvador. The people of Bahia and the city quickly began to lose their rural aspect. These spatial transformations and the city's way of life have been identified as collective social phenomena that have altered the city as a whole and its inhabitants. In *A Morte e a morte de Quincas Berro D'Agua,* these same phenomena are portrayed from Quincas' individual and ontological perspective: as his identity metamorphoses, revealing itself to another construction of his being, the city also becomes another. In other words, the city of Joaquim Soares da Cunha, a father of family and good manners, will not be the same as the Salvador of the "king of cachaga" known as Quincas Berro D'Agua.

Salvador and presented by Amado as the kingdom of Quincas. Amado plays with the dichotomy of bourgeois order versus the disorder of the scoundrels in his narrative. Quincas shows the conflict between these two social groups through a character who has moved from a middle-class neighborhood to the inner city of the "lumpen". In the city, especially in Pelourinho, Jorge Amado shows that there is interdependence, solidarity and spontaneity. The effective adjustment of these variables in the place offers the possibility of building citizenship, based on an endogenous process of character reconstruction. A good example and description of the city such as A Reign for Quincas is presented by Sellmann (2015):

"As they crossed the slopes of Sao Miguel, on their way to the castle, they were the target of various demonstrations. At the Flor de Sao Miguel, the German Hansen offered them a round of pinga. Further along, the Frenchman Verger handed out African amulets to the women. He couldn't keep them because he still had a saintly duty to fulfill that night. The doors of the castles opened again and women appeared at the windows and on the sidewalks. Wherever they went, they heard shouts calling Quincas, calling his name. He thanked them with his head, like a king returning to his kingdom". (page 84)

Quincas becomes the king of the wretched streets of Pelourinho, which also metamorphose into a vibrant kingdom. Amado plays with the multiple meanings of the word "castle", which means brothel in the popular Bahian vernacular language of that period. Prostitutes, in turn, become maidens in Quincas' kingdom. The procession of the lumpesinato also highlights other picturesque aspects: "the mixture of roles and the integration of differences." (Sellmann, 2013)

Salvador is also the city of carnival in *The Death and Death of Quincas Berro D'Agua.* Quincas, in his plunge into the "carnavalization" of the city/kingdom, literally takes dialogism to the streets in opposition to the monologic order of middle-class Itapagipe. However, even if celebrations like the one described above are as temporary as a Carnival party, dialogic life is permanent in the bars, brothels and residences of Pelourinho. This is the life that Quincas - once Joaquim - chose for himself. So his death had a great impact on the people who were part of his kingdom in the city of Salvador:

"The roda, in front of the sloop ramp, at the Agua dos Meninos night market on Saturdays, at Sete Portas, at the capoeira exhibitions on Estrada da Liberdade, was almost always numerous: sailors, small market traders, babalaos, capoeiristas, malandros took part in the long conversations, the adventures, the busy games of cards, the fishing under the moon, the sprees in the area.[...] Quincas' death increased the consumption of cachaga wherever it arrived. The news had spread faster than he had.[...] The rites of kindness of the people of Bahia, the poorest and the most civilized, were fulfilled. The mouths shut." (Amado, 1959, p. 64/65)

The carnivalesque aspect of chaos becomes a positive aspect of the city in the novel: an ideal element in contrast to the order established in the city's binary. Carnavalization can also be read in terms of the different processes of territorialization of the city: after ten o'clock at night, the city center welcomes prostitutes, vagrants and outcasts of all kinds, where they meet in poorly lit streets. This itinerant trade in fruit and edibles moves to the center under the gaze of customers in small fires lit on the sidewalks. The passer-by, still far away, smells the strong aroma of Afro-Brazilian delicacies, seasoned with dende oil and pepper, by black and mulatto women dressed in typical costumes. The bars become busy. The police relax their vigilance and the prostitutes can leave their homes and show off on the street. Quincas' journey reveals the city of Salvador in the novel: a binary represented by the middle class in Itapagipe and the lower class in Pelourinho. Quincas therefore acts as a guide to both places. To understand why Carnival is a favored element in the representation of the urban space in *The Death and Dying of Quincas Berro D'Agua,* it is necessary to contrast it with the order at the opposite pole of this society. (Sellmann, 2013, p. 173/174)

The duality existed in Quincas. However, his preference for the carnivalized life had long been decreed, much to the chagrin of his family. The choice was so clear that his daughter Vanda, even at the burial in the order of society that Quincas had abandoned, still maintained his decision after his death. On Vanda's feelings towards her father:

"He looked around for a place to sit. All that was there, apart from the cot, was an empty kerosene box. Vanda put it up, blew off the dust, sat down... [...] That father was a cross, he had turned their lives into a Calvary, they were now at the top of the hill, they had to be a little more patient. With the tail of her eye, Vanda peeked at the dead man. There he was smiling, finding the whole thing infinitely funny. [...] He had been waiting for them, his restlessness at the end of the afternoon was only due to the delay, the delay in the arrival of the tramps. Just as Vanda was coming to believe that her father had been defeated, that he was finally ready to surrender, to silence his lips with dirty words, defeated by her silent and dignified resistance to all his provocations, the smile shone again on his dead face, more than ever Quincas Berro D'agua was the corpse in front of her." (Amado, 1959, p. 75)

An ontologically striking aspect of Quincas' being in the historic center of Salvador is his social experimentation with malandragem. One of the strategies Amado chose to celebrate the magic of the place was to focus on disorder as a figurative element of Afro-Brazilian cultural practices. In *The Death and Dying of Quincas Berro D'Agua,* the malandro is someone who refuses to work regularly and makes a living taking advantage of other people, especially the rich. This exclusion from the job market is a way of life: "totally averse to work and highly individualized in his typical way of walking, his seductive way of speaking, and in his way of dressing" (Sellman, 2013, page: 119).

The permanence of the malandro in these narratives indicates a representation of Salvador in which alternative uses of the urban space become important. The malandro is associated with the lower-class areas of the novel. And the best way to understand this relationship between malandragem and the less privileged areas of the city comes from the "coronation" of Quincas' malandragem. Take a look at the following passage:

"He had spent ten years in this life: 'King of the vagabonds of Bahia', they wrote about him in the police columns of the gazettes, a type of street man quoted in the chronicles of literary men eager for the picturesque, ten years embarrassing his family. Splashing it with the mud of that undeniable celebrity. 'Salvador's most notorious cracker', o 'ragged philosopher of the Mercado ramp', 'the senator of the gafieiras', Quincas Berro D'agua, o 'vagabond par excellence', this is how he was treated in the newspapers, where his sordid photograph was sometimes printed." (Amado, 1959, p. 46)

Naughtiness will be portrayed in these different personifications that Quincas will carry. And all of them are intertwined in this "non-family", "non-middle class" space in the historic center of the lumpen city of Salvador. The malandragem play their part: to survive and reinvent their living spaces in contrast to the way in which these places were conceived. For Quincas, the life of the malandragem is synonymous with freedom, multiple possibilities and directions, a heterogeneity of his being, his ontological diversity. Meanwhile, middle-class life demands the sacrifice of all these things in order to maintain the homogeneous order.

The point of tension brought about by this issue in the novel comes from the perspective of Quincas' daughter, Vanda. She personifies the local differences in the book by comparing her father's behavior in Itapagipe in the "being of Joaquim" with the now verified "being of Quincas":

> How can a man, at the age of fifty, abandon his family, his home, the habits of a lifetime, his old acquaintances, to wander the streets, drink in cheap bars, frequent the street market, live dirty and bearded, live in an infamous pigsty, sleep on a miserable bed? Vanda could find no valid explanation (Amado, 1959, p. 24).

In the excerpt above, Vanda accidentally realizes that Quincas has become an extreme type of rascal, someone who literally abandons his home to live on the streets, becoming as satisfied as possible. Street life, which defies Vanda's sense of logic. Life on the streets - especially the streets of the city center - inverts the atmosphere of moderation that she finds at home. (Sellman, 2013, p. 178)

Having made this observation based on the contrast between Vanda and Quincas, it is important to point out another issue in this passage. The ontological changes born of Joaquim, now Quincas, are directly linked to the changes in his environment, the constitution of new geographies. It is in Pelourinho that Joaquim becomes Quincas, to lead a life that is the opposite of his middle-class routine, in a place that welcomes difference and conflict. Pelourinho provides a permanent dialog between opposites. In his journey to Pelourinho, the most resistant historical part of the city of Salvador, Joaquim will have the opportunity to change his identity from the fixed role of obedient employee to a multitude of other identities. Living in Pelourinho allows him to break free from the social bonds of the commercialized family. Frequenting the spaces of the historic center that is Pelourinho allows him to discover both his carnivalized and naughty side. There is no place in the city of Salvador where this tradition would be better represented and suited to the present day. To say where Quincas went, to look at the geographical landscape referred to the place where he lived and identified himself, also punctuates his relationships of belonging and what he has become. We will now analyze the construction of Quincas' geography as a legacy of his placement in Pelourinho.

3 - The Quincas Pillory:

Amado's text in *The Death and Dying of Quincas Berro D'Agua is* written about the "city of Bahia", which is how the writer referred to Salvador. However, he quickly locates the seductive mystery of the city and the polarizing center of Quincas and his work in the Pelourinho area. In this neighborhood, the binary city seems to dialogue more freely. This is especially true at night, when people who work nearby in the city center return to their homes in other neighborhoods to rest. The official city conventions, products of a synthetic form that is appreciated through class, are suspended to give way to a real dialogue when the city sleeps, and "the carnavalization of Pelourinho reaches its peak". (Sellman, 2013, p. 181)

Amado's proposal for examining the heterogeneity of social interactions and places in the historic city center of Salvador comes from the opening of the plot with the death of Quincas. Amado provides a guide to the area when in the novel the narrator recounts the reactions to the news of Quincas' original - or second - death. Take a look at this passage:

"There were hasty gatherings in the vicinity of the Market, like lightning rallies, people walking from one side to the other, the news in the air, going up the Lacerda Elevator, traveling on the streetcars to Calgada, going by bus to Feira de Santana.....] On the sloops with their sails down, the men from the kingdom of Lemanja, the tanned sailors, couldn't hide their disappointed surprise: how could this death have happened in a room in Tabuao, сото the 'old sailor' had died in a house? Hadn't Quincas Berro D'agua, with a voice and manner capable of convincing even the most unbelieving, proclaimed peremptorily that he would never die on land, that only a tomb was worthy of his piety: the sea bathed in moonlight, the waters without fire?" (Amado, 1959, p. 53/54)

Quincas' death reconfigured some of the usual dynamics of the Pelourinho portrayed by Jorge Amado. And in a rather picaresque way, just like the dead man's choice of daily life. Tourist stores in the vicinity raised the nails of their souvenirs to pay homage to the deceased. People commemorated the dead man in "bars, botequins, on the balconies of sales and warehouses, wherever cachaga was drunk" (Amado, 1959, p. 50).

The whole of Pelourinho, with its colonial physiognomy, the houses, churches (Nossa Senhora do Rosario dos Pretos, Ordem Terceira do Carmo and Convento do Carmo), Cruz do Pascoal, Beco do Mota, Ladeira de Sao Miguel and many other places, make up the landscape that serves as the setting for this picturesque story. The urban site of Pelourinho, whose listed area constitutes, "in the heart of Salvador, a small city within the city", represents a cultural heritage site (Carneiro, 2004).

Disorder is also an important feature of the Pelourinho environment.

In this geographic environment, geographies meet: central markets like the Agua dos Meninos Market are both workplaces and spaces for leisure and socializing. Disorder can be read both as urban chaos and as resistance to a hegemonic order of social life and the commodified axes of everyday life. The following passages demonstrate this "disorder" regarding the environment that is Pelourinho when Quincas' death is described and located there:

"Ladeira do Tabuao was no place where a lady could be seen at night, a street of ill repute, populated by scoundrels and women of life. The next morning, the whole family would gather for the funeral [...] the scoundrels who told Quincas' final moments in the streets and hillsides, in front of the Market and at the Agua dos Meninos Market, were disrespecting the memory of the dead man, according to the family. And the memory of the dead, as we know, is a sacred thing, not to be in the unclean mouths of potheads, gamblers and marijuana smugglers." (Amado, 1959, p. 18)

"The group laughed; in the bars the noise started up again, life had returned to the Ladeira de Sao Miguel.[...] As they crossed the Ladeira, on their way to the castle, they were the target of various demonstrations. At the 'Flor de Sao Miguel', the German Hansen offered them a round of pinga. Later on, the Frenchman Verger handed out African amulets to the women. The doors of the castles opened again, and women appeared in the windows and on the sidewalks. Wherever they went, there were shouts calling Quincas, calling his name and him again. He thanked them with his head, like a king returning to his kingdom." (Amado, 1959, p. 96)

The first passage shows the view of Pelourinho from the perspective of Joaquim's family. This is a place that they only had to visit momentarily because of the character's death. The regulars and friends of Joaquim/Quincas are, according to the family, disrespectful to their relative. However, in the second passage, we see Pelourinho as Quincas' enchanted kingdom. His space of ontological transformation, where, as described, the doors of this castle allowed him not only to enter another environment, but also to become another man. Thus, definitively ending his time circulating on the slopes of Pelourinho, as well as among the people who lived there, was his best way of feeling alive one last time. Quincas' scenario was not only building him, but also making him be once again.

From this historic center of the city of Salvador, a spatial translation of this disorder is presented: its locally settled inhabitants, different people, such as the aforementioned prostitutes, scoundrels and vagabonds lived degrading lives in the eyes of the middle class. Disorder, as we will see in the Amadian novels, not only meant misery and decadence, but also an unwanted confrontation with differences.

Amado's book naively glorifies the naughty/carnival life in Pelourinho. However, it also sheds light on something else: the multiplicity of uses that transforms Pelourinho into a dynamic space despite all the poverty and squalor. Commercial houses share the area with residences and churches; the proximity of different cultural traditions to negotiate and tolerate - if not accept - differences. Public spaces become extensions of private spaces, even in extreme ways, such as "Quincas' habit of sleeping anywhere he chooses" (Sellman, 2013, pp.182/183).

Even accepting that poverty is romanticized in *The Death and Death of Quincas Berro D'Agua*, Pelourinho still offers models of use and interaction for an ideal urban space. There is a constant interactive flow in the city's historic center, the "intense life" that Amado actively verbalized in the book. In spatial terms, there is also an intense exchange of functions, as commercial spaces share residential areas, and they can even share the same buildings. Some of these places have a dual function: work and socializing in the botequins and casas de Mulheres de Sao Miguel, which resemble the markets mentioned above. The city's historic center is characterized by multiple uses of space (Sellman, 2013, p.184).

The announcement of Quincas' second death at a certain point in the book affects the daily spatial structures of Pelourinho and the people who lived there:

"Also in those poor houses of the cheapest women, where vagabonds and scoundrels, small-time smugglers and disembarked sailors found a place, family, and love in the lost hours of the night, after the sad market of sex, when the weary women longed for a little tenderness, the news of Quincas Berro D'agua's death was heartbreaking and made the saddest tears flow[....] In the late afternoon, when the lights were coming on in the city and the men were leaving work, Quincas Berro D'agua's four closest friends - Curio, Negro Pastinha, Cabo Martim and Pe de Vento - were walking down Ladeira do Tabuao towards the dead man's room. They looked longingly at the market, the trucks and sailors in the street, the canoes on the sea, the people coming and going. They felt a sudden emptiness, they couldn't even hear the birds in the cages nearby, in the stall of a market vendor." (Amado, 1959, p. 71)

Instead of being places that generate insecurity and fear, Pelourinho's public spaces are seen as a medium that provides opportunities for socializing. Streets and sidewalks can provide the settings for spontaneous celebration. Residents of the historic center of the naughty/carnival city of Salvador see the streets as a place for more stable interaction and more time. The streets are part of this life, which redefines and reconstructs beings:

"It looked like it was going to be a memorable, unforgettable night; Quincas Berro D'agua was on one of his best days. An unusual enthusiasm had taken hold of the group; they felt they were the masters of that fantastic night, when the full moon enveloped the mysterious city of Bahia. On Ladeira do Pelourinho, couples hid in the centuries-old portals, cats meowed on the roofs, guitars wailed serenades. It was a night of enchantment, the sound of atabaques resounded in the distance, the Pelourinho looked like a ghostly scene." (Amado, 1959, p. 93)

Several works outside of literature have also been written in which Pelourinho is a vernacular means of geography. The geographer Milton Santos, for example, published his master's thesis on the center of Salvador in 1959. This is how Santos described Pelourinho:

"The Pelourinho is an irregularly shaped hillside, surrounded by 18th and 19th century buildings, large two and three-storey noble houses that once served as residences for wealthy families, but which today are now ruins. [...] The ground floor of all these buildings is occupied by shops and handicrafts. [...] On the ground floors, a heterogeneous population lives in more than precarious conditions. [...] Damaged staircases, cracked floors, dirty walls and leaking ceilings form a common picture throughout this area of degradation." (Santos, 1959, p. 117)

This stretch of Santos stands in stark contrast to the "disorder" narrated by Jorge Amado in *The Death and Death of Quincas Berro D'Agua*. Spaces with multiple uses. Coexistence, even if marked by conflicts of difference between different social classes. Socialization and precariousness in the coexistence between people. The heterogeneity of experiences. From this logic, where the Pelourinho is Quincas' means of transformation, permeating his symbiosis of different classes from different moments in his life, his dualities are also found: the conflict of classes. At times he is Joaquim through the eyes of his family visiting Pelourinho. At other times, he is Quincas, the king of this neighborhood who, in addition to making order complex, breaks with the social bonds of the middle class. We'll set out to problematize this issue.

4 - Joaquim's middle class or Quincas' lower class?

Some subjects in Jorge Amado's works have gained an established status. One of these is the debate about relations between different social classes. Or rather, the clash witnessed in everyday life, in the space-time dimension very well represented both in the countryside and in the cities. The writer never hid the way he problematized this issue: from a dualistic and Manichean perspective. The binary "poor (lower class) and rich (middle or upper class)" defines social positions in opposition, waging a conflict in a space of struggles. This is undoubtedly another legacy of the communist ideology that permeated the author's thoughts for a long period of his life. The Manichaean bias was the very inspiration for the social realist novel, in which the Soviet propaganda issues of the time were problematized by the characters. This Manichaeism was given through the representative gaze of the classes: the poor were the "good" and persecuted, the most unjust, and the rich and/or middle class were the "bad" exploiters and privileged.

The great novelty in *Death and the Death of Quincas Berro D'Agua* is that this Manichaeism is a little more masked in the subjective and ontological questioning of the protagonist Quincas Berro D'Agua's passage through life. Amado chooses to represent the proletarian novel less and works on class issues in a more internalized way with the protagonist: choosing to live according to one of the classes in this Amadian binary system and also deciding what kind of man he is going to be and build himself. Quincas's choices throughout the plot build up to the moments represented by "his deaths". Take a look at the following passage:

"[...JQuincas Berro D'agua, when he died, was once again that old and respectable Joaquim Soares da Cunha, from a good family, an exemplary employee of the State Revenue Office, with a measured step, a scruffy beard, a black alpaca jacket, a briefcase under his arm, listened to with respect by his neighbors, giving his opinion on the weather and politics, never seen in a bar, with a homemade and measured cachaga. In fact, in an effort worthy of all applause, the family had managed to make him shine." (Amado, 1959, p. 20/21)

In *The Death and Dying of Quincas Berro D'Agua,* different perspectives permeate the narrative. There are two versions of Quincas Berro D'agua's story: one told by his middle-class family, and the other by his friends from the lower classes of Pelourinho. The narrator tries to maintain a balance by refusing to acknowledge that either of them is the more faithful version of events. "To this day there remains some confusion about the death of Quincas Berro D'agua," is how Amado opens the narrative of his book. Salvador seems to be divided into two sections in the narrative: the petty bourgeois community of Itapagipe, which follows the order of the capitalist system closely, and the proletarians and scoundrels of Pelourinho, who live in carnival chaos. Each sector has its own narrative focus. Using ambiguity throughout, Amado maintains the conflict between the versions and places until the end of the story.

There are two perceptions of spatial confrontations in Amado's geographical environment in these narratives: one by a repressed middle class and the ruling elites, and the other by the lower classes. However, if Amado satirizes the established order, with Quincas' choices he also proposes an alternative to it in the way he did in his proletarian novels?

Joaquim dies and becomes Quincas, as Amado tells us. However, Quincas had also been Joaquim, a total bohemian who in another time in Pelourinho had once been a dedicated employee of

Itapagipe; a middle-class citizen who chose to live among the Soteropolitan scoundrels of the center; and, finally, the choice the writer makes is to read the man who crosses worlds, neither alive nor dead, or perhaps both. Duplicity and ambiguity are Amado's choices. He is only concerned with defining that each of the protagonist's lives is tied to different times and spaces. On the first look of the protagonist's daughter Vanda at his death:

"[...] He was the corpse of Quincas Berro D'agua, a coward, a debauchee and a gambler, with no family, no home, no flowers and no prayers. He wasn't Joaquim Soares da Cunha, a correct employee of the State Revenue Office, retired after twenty-five years of good and loyal service, a model husband, whom everyone took off their hats and shook hands with." (Amado, 1959, p. 27)

Even after the preparation of Quincas' body at Joaquim's, Vanda looks at her father and notices his box:

"He saw the smile. A cynical, immoral smile, the smile of someone who was enjoying himself. The smile hadn't changed, the experts at the funeral parlor had achieved nothing against it.[...] It was still that smile of Quincas Berro D'agua and, in the face of that smile of mould and enjoyment, what good were new shoes,[...] what good were black clothes, a clean shirt, a clean shave, starched hair, hands folded in prayer? Because Quincas laughed at it all, a laugh that grew wider and wider, that gradually resounded in the filthy pigsty. He laughed with his lips and his eyes." (Amado, 1959, page 50)

Vanda, Quincas' daughter, is the personification of the middle class. In *The Death and Dying of Quincas Berro D'Agua,* the middle class is portrayed as the conservative class, which abhors the lower class. His main indignation within this problematized logic is the fact that his father - according to his existential order - was a "good man" and then chose to be a "rascal" on the streets of Pelourinho. However, as Quincas embodies human misery itself, stratified citizenship is evident. Being a type of citizen is defined through the construction of their geographies associated with which part of the geographical environment of Salvador they live in, in the case of the narrative.

It is important to note, as we have already mentioned here, that Salvador has undergone processes of modernization that have resulted in different logics of re-signification of its daily life in the geographical environment and the spatial-temporal practices of its subjects. However, the historic central area of Pelourinho continued to be a poor enclave of prostitutes and various types of precarious subjects, but its landscape of historic buildings was reappropriated for new uses, and also gained status as a tourist attraction, feeding a new sector of the economy. In order to better visualize the binary logic, this modernization that the city went through in a certain period is also a relevant analytical tool for us. The city's economic development unfolded into the growth of a middle class that was eager to experience the latest trends in bourgeois life in the capital of Bahia (Sellman, 2013, p.167).

Vanda and her family strictly followed a concept of time that is subject to the capitalist order - the time of the economic space, workplaces and businesses - which dictates that "time is money".

Joaquim's family believed that prosperity came from social and professional recognition. Thus, in his memoirs about his father, he was the employee honored by his colleagues. He was the family's financial provider. At a specific event narrated in the book, the family home was packed with middle-class people to pay tribute to his father: he had been promoted in his job. His aspirations for happiness in the value of the urban space fit into his quest to own a property. From her father's point of view, that tribute was the realization of his existential prison in a life that he was beginning to realize he wanted to overcome. At the tribute, which Vanda hadn't forgotten at the time, "Joaquim listened to the speeches, shook hands, received the pen without showing any enthusiasm. Joaquim wanted to escape this commercialized reality" (Sellman, 2013, p. 177).

Thus, when Joaquim chooses to reconstruct his existential reality, his first choice is to change his place. The geographies that result from the ontological construction of his being necessarily involve place. Since the search is for a new existence, this change is essential. From here, his spiral center leaves Itapagipe and goes to Pelourinho. From the middle-class Joaquim of Itapagipe to the lower-class Quincas of Pelourinho.

In this change of environment for the protagonist - and thus the birth of Quincas - Amado plays with the multiplicity of beings that the character becomes. Amado plays with Quincas' double role, as well as with the uncertainties surrounding his death. Quincas paradoxically becomes a "king":

"The local newspapers alternately portrayed Quincas as the 'king of the vagabonds of Bahia', the 'ragged philosopher of the Market Ramp', and the senator of the gafieiras" (Amado, 1959, p. 39).

These carnivalesque hierarchies and descriptions become respected occupations in the identification of Quincas, the roles ironically bringing respectable titles to street level. The historic center of the city, which Vanda perceives as threatening, becomes magical for Quincas who dies Joaquim and is reborn. Quincas literally brings dialogism to the streets in opposition to the monologic order of middle-class Itapagipe. However, even if celebrations like the one described above are as temporary as a Carnival party, "dialogic life is permanent in the bars, brothels and residences of Pelourinho." (Sellman, 2013, p. 180).

Quincas' new home is the Pelourinho of Salvador's lower classes. Afro-space, a space of entertainment and also of habitation. Disorder is also an important feature of the Pelourinho environment. However, it's important to ask: under whose logic? On the other hand, the diversity of life in the poor streets of Pelourinho can't keep the confrontation at bay with that order now under the prism of the middle class that Quincas leaves behind for new definitions of his being. It's a geographically complex environment for which, in his opinion, his freedom was born. In the midst of the problematic glorification of the precarious social being that Quincas chooses for his life, the novel makes a case for disordered geographical settings that allow for multiple uses, which break with their original conceptualization. Thus, the protagonist remakes himself. Joaquim or Quincas, by transforming themselves, re-existing within the ambit of their "deaths", reconfiguring their geographies, dialectically turn their scenery into a new geographical environment.

5 - The geographies *of Death and the Death of Quincas Berro D'Agua:*

The choice of *The Death and Dying of Quincas Berro D'Agua* as the book to close this final chapter of the thesis was not random. Many of the geographical issues that have already been addressed throughout this research - such as the problems of the geographical environment, the emergence of geographies and ontological readings of an urban space presented in a literary work - are empirically repeated in this book. However, the existential construction of the protagonist in his environment(s) denotes an even greater reinforcement of what has already been problematized throughout the thesis text.

Many of the different geographical approaches already developed in the previous chapters would fit perfectly both for this chapter, which refers to Quincas' book, and for the conclusion of this thesis. It is important to emphasize that we believe that an epistemological debate that has already been built up in this research about the other previous books in their respective chapters is also possible for *The Death and Dying of Quincas Berro D'Agua.* It is also important to state that the same categories chosen as theoretical propositions in the thesis - such as geographical environment and geographicity - continue to permeate the epistemological reading of the thesis in this final chapter. However, with a view to a conclusive path, the great difference of this book lies in the more subjective/introspective debate that is proposed in the Amadian narrative of Quincas.

The existential setting of the narrative is once again the city of Salvador. And present in the same existential context as the protagonist is the possibility of differentiating him as a being through the part of the city in which he is perceived in the plot. The city center of Salvador is once again portrayed as the city of Bahia itself. The city of Salvador is the oldest and most characteristic of Brazilian cities. For more than half of Brazil's history, it was the most important urban agglomeration in the country. However, during the 20th century, the axis of the Brazilian economy, which had its nerve center there, shifted to the Southeast. This situation had a profound impact on the city's physiognomy and on the lives of the people who were born and live there, especially in this central part of the city, which is so well represented in the works of Jorge Amado.

In *The Death and Dying of Quincas Berro D'Agua,* by referring to an individual micro-scale of the life constructed in Amado's narrative of his protagonist, we can allude to the very changes that have so transformed the geographical dynamics of Salvador. Before Joaquim from the middle class of Itapagipe, then Quincas from the *lumpen* of Pelourinho. Normatized by a hegemonic order, we could compare the prosperous Salvador of an oligarchic Brazil still in the colonial structural molds, to the current stagnant and recovering Salvador, albeit with secular marks on its landscape. However, the most interesting question to ask about these reflections and allusions is: whose referendum is the order? Amado seems not only to invert the logic, but also to reflect on the established alliances of a hegemonic economic proposal.

Associated with this reflection, we will propose another epistemological discussion in parallel to finalize the problematics of the thesis: what is the landscape we are dealing with? Landscape, this concept so geographically marked in a scientific way, will also be problematized in its meaning here. For these reflections, we will work with the ideas of Besse (2006) and Cauquelin (2007) on the concept of landscape; and Santos (1958) in his thesis on the city center of Salvador.

5.1 - The landscape and the landscape in The Death and Death of Quincas Berro D'Agua:

The story of Quincas Berro D'Agua is set in the center of the city of Salvador. It is the center of a large city - Salvador - and therefore the founding theatrical setting of the plot in its existence. This terminology has already been mentioned earlier in this thesis, but it bears repeating here to reinforce the importance of this idea. The synthesis of this ontological theatrical scenario is manifested by the creation of a landscape. According to Santos,

"The landscape, then, is the result of a combination of elements whose dosage supposes a certain rhythm of evolution and a certain dynamism and the element of contradiction is represented by the factors of inertia" (1958).

However dynamic and entrepreneurial the city centers have become structurally, they still have an air of family, which comes from the functional historical concentration of other times. In general, this is the case of Salvador and the big cities that mark a kind of link between a rural world, over which life presides, and another world that is modern according to urban-industrial logic. In *The Death and Dying of Quincas Berro D'Agua,* forces of transformation and resistance clash in the shaping of the character. This tension is represented through his "deaths". In each of these "funereal" moments of Quincas in the narrative, we see the result of the creation of an entirely new landscape, either the transformation or adaptation of the old landscape, which then degrades or retreats. Wouldn't it be possible to make exactly the same reading of change in Salvador's city center? Take a look at the following passage from Quincas' narrative:

"The family council didn't last long. They argued at the table of a restaurant in Baixa do Sapateiro. The crowds were hurrying down the busy street. Across the street, there was a movie theater. The corpse had been entrusted to the care of a funeral parlor.[...] In the expert hands of the funeral parlor, Quincas Berro D'agua was becoming Joaquim Soares da Cunha again, while his relatives ate fish stew in the restaurant and discussed the burial."

(Amado, 1959, p. 35/36)

This passage refers to the initial presentation of the context of the plot of Quincas Berro D'Agua. The constitution of the landscape is mixed with the presentation of Quincas' first death in the novel. The concept of the city of Salvador passes through the drama of Quincas' first death, but at the same time it is also concretized in that moment of family reunion at specific points in Pelourinho, the city center of Salvador. The landscape is the backdrop of the plot and is also what constitutes its main points of definition.

With a view to further empiricizing this theoretical proposition, in another passage from Amado's book we see the natural, architectural and social elements of the city blending with Quincas' daughter's anguished conclusions about the moment of her father's death:

"The heat was rising in the room. Once the window was closed, there was no sea breeze to enter. Vanda didn't want it either: the sea, the port and the breeze, the slopes climbing up the mountain, the noises of the street, were all part of that finished existence of infamous madness. There should only be her, her dead father, the late Joaquim Soares da Cunha and the most cherished memories left by him." (Amado, 1959, page 48)

Studying a city's landscape geographically through literary narrative, even before analyzing its specific problems, can lead to a series of more general questions about the epistemological value of this proposal. Could this landscape narrated by Amado be the constitution of a geographical reality? We believe not only that it is possible, but that this type of reading would pay attention to existential minutiae that might not be very clearly demonstrated in established studies of urban geography, for example. The individuality of epistemological proposals carries individualities that lead us to distinguish cities into various groups. In this reading, the city of Joaquim/Quincas Berro D'Agua. However, this view gains in relevance because it could perfectly well apply to any other citizen of this place. It would be enough to have experienced the same ontological anguish as Amado.

The use of landscape as a geographical concept is accompanied by another issue: the spatial area in question. Here, it is worth reflecting on the scope of this geographical environment in which the city center of Salvador is defined in this chapter. In other words, dealing with geographical landscape is a problematization that involves scale. In geography, landscape is apprehended by the human senses, especially sight. However, when scientifically problematizing artistic works such as a painting or a book, the notion of scale can be less rigid than a mathematical cut-out.

Cauquelin, in his publication "The Invention of Landscape" (2007), which will be our reference here, proposes an analysis of how the concept of landscape emerged and became epistemologically established. Cauquelin argues that landscape is a substance, where it is continually confronted with an essentialism that transforms it into a natural fact. Through this reading, landscape can be confused with nature itself, something that is not part of our analysis here. Cauquelin (2007) argues that the landscape is endowed with symbolism. It is by agreeing with this proposition that we can think even more clearly about the landscapes of Quincas and Amado. She says of this reasoning:

"Here, an ontology is summoned that renders all discussion of a probable genesis moot. That the symbolic form 'landscape' was formed over centuries is therefore inadmissible, because if landscape is identified with nature, it has always been there." (2007, p. 39)

According to this idea, it is necessary to discard the intention of trying to find the emergence of both the landscape and its conceptualization. If landscape is identified with the natural environment, it is an inexorable fact of the world's geographies. On the other hand, if landscape is also substance, it is in a constant state of becoming. The construction of the landscape is to be found before man and undoubtedly after him. The geographical landscape proposed here is that of eternal construction. Joaquim de Itapagipe is the man of a landscape. When he leaves his family behind in order to reduce his existential anguish, he dies and Quincas announces himself, moving to another landscape. And we see his final geography in this clash: in his natural physical death, what will be the outcome of his existence, the man of the lumpen landscape or the one who returns to the bosom of his family.

Landscapes can be constructed and revealed to us on a daily basis through the artifice of technology. This seems to be the most common resource in the apprehension of landscapes in large cities, such as Salvador. In the modern city, roads and expressways, bridges and streets, plagues and open spaces transform our uses, free or hinder walking, provoke some of our habitual gestures and condemn others. The landscape of the city is the one that most reminds us of movement, of displacement. What's even more interesting is that this movement and/or displacement of people through cities, when mapped in the landscape, also defines them. Take this passage from Amado's book about a friend of Quincas:

"Cabo Martim could be in three or four places. Either sleeping at Carmela's house, still tired from the night before, or chatting on the Market ramp, or playing at the Agua dos Meninos Market. Martim had devoted himself to only these three occupations since his discharge from the army some fifteen years earlier: love, conversation and gambling." (Amado, 1959, p. 66)

It's interesting to see in this passage how much the displacement of the character Martim is visually represented in the narrative. Amado's words seem to convey quite clearly the mental construction of an image of the landscape. And, in order to use a literary work to problematize the geographical landscape we are considering here, this constructive aspect of the plot is essential. Cauquelin argues that

"the landscape that stands before me and offers me its proposal fills in the conditions of its production between the spectacle before me and the general form in which it must move so that I can apprehend it" (2007, p. 118).

This construction of the image of the landscape through the narration of the book's plot can offer us a dialectical methodology. This path is perceptible when there is the necessary transformation of reality into image and, again, of image into reality: in this double movement, something, a breath is transmitted. Because, turned around, reality is no longer exactly the same: it is duplicated, reinforced by fiction. In this sense, the scientific basis of the relationship between Geography and Literature is further strengthened.

Often in the case of landscape, this argument from dialectical methodology ends up explaining why nature is not synonymous with this concept. Our mental functioning agrees and resonates with this same construction: nature, pure exteriority, also becomes pure interiority. We have the intimate feeling of perfection, of a nature-to-nature relationship. For Cauquelin, this event stems from a double guarantee: "nature (exterior) guarantees the landscape, and the landscape guarantees - acts as guarantor - of the naturalness of our nature (interior)" (2007, p. 124).

Landscape is not synonymous with nature. However, we social beings are integral and founding elements of it. And on this point, our geographies emerge and this cannot be forgotten epistemologically. About this:

"Nature is the indistinct whole, of which we are an intimate part, but a part that is the whole. Totally implicit, totally evident, without the shadow of a question about its fabrication, the perfect landscape emerged in the universe of elementary forces. It is the beginning of the world and its end. What remains last when, on all sides, certainties crumble is precisely this affirmation of itself as nature and the feeling of an agreement that has nothing conventional about it." (Cauquelin, 2007, p.125/126)

In a passage from *The Death and Dying of Quincas Berro D'Agua,* it becomes quite clear how much this reading that we are also nature marks the ontological definition of the protagonist and even of other Amadian characters:

"On the docks and beaches, children were born knowing the things of the sea, there's no point in looking for explanations for such mysteries. Then Quincas Berro D'agua made his solemn vow: he would reserve for the sea the honor of his final hour, his final moment. They wouldn't imprison him on seven feet of land, no! When the time came, he would demand the freedom of the sea, the journeys he had not made in life, the most daring crossings, the deeds without example." (Amado, 1959, p. 57)

Another interesting issue to reflect on is that landscape is not a pure synonym for nature, but rather the method by which we apprehend the contents of the geographical landscape. As well as the method itself, the meaning of landscape includes the technique used to observe it. Cauquelin (2007) once again gives us an interesting metaphor for analyzing the landscape. She says that the "window" - of the eye or of the soul - is a landscape instrument par excellence. But what is the window beyond its meaning as an object? The writer also places the idea of the window as something intrinsic to our subjective capacity to absorb the world. Her words:

"[...] it is also the soul, whose window is the eye, which governs vision. There is no doubt that here we have a sine qua non condition: the window and the frame are 'passages' to the vision, to see landscape where, without them, there would only be... nature." (Cauquelin, 2007, p. 137)

We will therefore read landscape as the culturally instituted presentation of that nature which surrounds us and also dialectically grounds us. There is a landscape in the Salvador of Amado and Quincas. In turn, just as it has been building the protagonist in his wanderings, his displacement has been redesigning this city.

It is also worth reflecting on what we would understand as the urban landscape of the city of Salvador in *The Death and Death of Quincas Berro D'Agua.* All the more so because throughout the thesis we have chosen Amado's books with a city theme. Urban landscape is an expression that seems to contradict the natural notion of landscape, both because it denies the very close relationship between landscape and nature, and because of the content, often potentially artificialized and technical, offered by the vision of a city erected in disparate towers, pierced by vacant lots, with wrinkles of historical architecture, saturated with dirt and bathed in the opaque smoke of industrial objects. Yet, even in a picture that seems adverse, we see this spectacle сото landscape. The city is also framed by the "window":

"We frame, we make the city into a landscape through the window we interpose between its form and us. Numerous *fences,* a street corner, a window, an open balcony, the perspective of an avenue. The prospect here is permanent. The city participates in the very form of perspective that produces the landscape. It is, by its very origin, nature in landscape form." (Cauquelin, 2007, p. 149)

Both in the dialectical sense and in the substantial sense of the city, its landscape is more clearly landscape than the harsh and/or natural landscape. Its construction is more marked, more constant. The city is at the heart of the shifting movements of people's being. There is no doubt that it is through this bias that nature is present in the landscape, not because it would be a part of it, valid for the whole, but because it is produced by a sequence of rules, whose coherence makes an object in everything and by everything similar to a natural object.

Taking another scalar perspective, it is also worth analyzing the concept of landscape in the light of Besse's (2006) problematizations. Landscape, as previously mentioned, can be worked on scientifically in Geography under the aegis of different scales. In his text, Besse develops a more comprehensive approach, in which the Earth is seen as the scale par excellence of the landscape. However, he also allows the assertion that the scales of the landscape are multiple: the province, the country or even the region. Although in a different spatial context from our problematization in this chapter, some considerations about his theories are also pertinent.

The first important point to reflect on in Besse's (2006) reasoning is the space of the landscape as an objectification of existence. Affirming this type of proposition contributes to our understanding of geographical epistemology, where landscape is an important concept for the theorization of our science. According to Besse,

"More precisely, the landscape visually and imaginatively translates the promotion of geography as a specific discourse, distinct from cosmography, dedicated to the description of the universal Earth." (2006, p. 23)

The point of convergence between Cauquelin's (2007) and Besse's (2006) theoretical reading of landscape lies in the fact that they use artistic manifestations such as painting and literature as perfect arrangements and/or instruments for conceptualizing landscape in Geography. In both cases, literature - which is the manifestation that is pertinent to us - is a way of evoking the very history of human formation. In our research, Salvador is perfectly seen, from a Bessian point of view, from a landscape reading. Just like the existential construction of the protagonist of the work. Take a look at the following passage on the reading of landscape through painting in Besse's text:

"What is unique about the paintings that represent such 'landscapes of the world' is this way of linking and encompassing the accidents of space (trees, rocks, buildings, rivers) in a unity that develops from the background, an indefinitely open background that refers to a cosmic space and time within which human history is as if evoked in its relativity." (2006, p. 25)

The metaphor of the theatrical stage as the founding ontological element of man also appears in Bessian theories, but in a reading of the Earth's landscape as a whole. The scale is different, but the theoretical foundation is the same as that already discussed here. He states:

"It's a new kind of experience of the Earth that seeks the means for its formulation in the piano of representations and discourses, often using the universe of ancient models. This structure of perception and thought is theater." (Besse, 2006, p. 29)

But what is the Earth in this theater? And how is it perceived? Let's emphasize this point: the Earth in the map and landscape painting that represent it becomes an object for a subject who is its spectator, just as this same spectator constructs the substance of the landscape. Corroborating this, let's think about what the city of Salvador is in this theater? What is the city center of Pelourinho in this context? How is it perceived? Under the logic of the world and the gaze of Quincas, a protagonist who dies to rebuild himself.

The Earth or the center of Salvador no longer crystallize time, they become the frame, the support, the theater of its unfolding. Time and history have become spectacles. Amado's narrative is also its own handle filled with geographies. The geographical environment is not just the inscription of time, nor is the Earth the result of genesis. They are the conditions within which the world develops its meanings.

Developing this idea of landscape as theater, Besse (2006) states:

"A constitutive paradox, no doubt, but one through which the Earth and man receive their true status, on the one hand an image, on the other the one who contemplates it. A paradoxical structure of subject and object, which is summed up in the metaphor of the theater, a theater in which the human being is both actor and spectator, at the same time inside and outside the scene, considering it сото an image." (page 29/30)

Thus, this theatrical device, in which the Earth is presented as an image and man is seen as both a participant in and an outsider to the scene, also manifests itself in *The Death and Dying of Quincas Berro D'Agua* and in the other Amadian texts discussed here. From the elucidation of this passage above, once again the ideas of Besse and Cauquelin problematized in this chapter come together. We can also see in this passage by the writer mentioned above a dialectical methodological proposition, in which landscape-meaning-human are re-founded.

Continuing from a Bessian perspective, landscape is the order of the world that becomes visible. Consequently, the birth of landscape raises this question: what does it mean that an element that was originally translated exclusively by theory now requires an aesthetic representation? Or even, what will be the function of the landscape in relation to the human concern to draw a totalizing horizon for its existence, that is, a world? Says Besse (2006):

"However, the landscape is not only the place of this very special 'pleasure' that is aesthetic pleasure; it possesses an irreplaceable cosmological and ontological density which, moreover, guarantees aesthetic pleasure a specific vocation." (page 35)

By stating that the landscape must have an aesthetic sense of vocation, or even ontological density, once again Bessian ideas confirm the theoretical propositions of the thesis. Even if Literature represents a stylistic whole in the content of its work, once the geographical environment of the plot interferes in the existential formation of its characters, an ontological perspective is born. And, in Amado's case, characters who to this day walk the streets of Salvador. In the following passage from Amado's book, we don't need the historical and temporal context to perceive the approximation of something almost identical that can still occur today in the city of Salvador:

"The santeiro, a thin old man with a white face, went into great detail: a black woman, who sold porridge, acaraje, abara and other delicacies, had an important matter to discuss with Quincas that morning. He had promised her to get some hard-to-find herbs, essential for candomble obligations. The black woman had come for the herbs, it was urgent to receive them, they were in the sacred season of Xango's feasts." (Amado, 1959, page: 22)

It is possible to propose the following hypothesis regarding the function of landscape representation and its relationship with geography: the landscape highlights what geography is about, that is, the sensitive experience of the Earth as an open space, a space to be traversed and discovered. More broadly, the representation of the landscape graphically embodies the new thinking and the new experience of the Earth as the universal ground of human existence. These are ideas from Besse (2006), which perfectly fit the following passage from Amado's book:

"The moon rose over the city and the waters, the moon of Bahia, in its silver sprawl, came through the window. The sea wind came with it, blew out the sails, the box could no longer be seen. The melody of guitars drifted down the slope, a woman's voice singing feathers of love. Cabo Martim began to sing too." (Amado, 1959, p. 88)

Here, the description of the landscape was found on different scales: that of the Earth with the elements of nature. That of the city of Salvador with the presence of the sea. That of the city center with the Pelourinho and its hillsides. And finally that of Quincas, in the sea wind blowing out the sails of his box. The landscape is thus, and let's insist on this point, not only the extension of the vocabulary, but also the visual illustration of the new geographical experience of the world.

5.2 - A geographical look at the landscape of downtown Salvador:

Salvador offers those who reach it by sea a glimpse of the spectacle of an Upper City built on hills and valleys, and a Lower City on a narrow plain. It is no coincidence that this is how the central part of the city is popularly divided. This is the designation that retains all the interest of understanding the central part of the city. The urban landscape of the city of Salvador - as it has already been understood here - is made up of wide, straight avenues, on flat surfaces conquered from the sea. It's notable for the presence of both luxurious newly-built properties and old dilapidated houses in narrow, winding streets. This neighborhood of contrasts found as the mark of four centuries of construction endures to this day. For some, all this apparent disorder is what gives this part of the city its beauty and color.

Death and the Death of Quincas Berro D'Agua was a book written at the end of the 1950s. Jorge Amado was already aware of these particularities about the city. However, the notion of the city's growth and the formation of the periphery did not have the scope of the present day. In his book "0 centro da cidade de Salvador", Milton Santos (1958) published an account of the urban revolution in a time frame corresponding to Amado's Salvador represented in this book, which is now being analyzed. Thus, combining this information does not seem meaningless or mistaken. One point to think about with regard to Salvador, as put forward by Santos (1958), is the combination of different historical periods that can be perceived in the city's landscape. Take a look at this passage:

"The variety of streets, the generations of buildings, mean much more than the urban or architectural preferences of one era or another: they are the mosaic of the centuries, but they also represent the succession of techniques, the whole evolution of urban life, the sum of the past and modern ways of being, whose incorporation into urban life does not always follow the same rhythm." (Santos, 1958, p. 103)

Milton Santos' (1958) reading of Salvador's city center is quite technical in relation to the existential analysis being developed for the geographic issues of this thesis so far. However, it is important to understand in his reading the relationship between time and people's way of being, filling the landscapes and also shaping the geographical environment as we saw in Jorge Amado's work.

The 20th century, which is the time of the books by Amado and Santos, was the time of the great transformations of the center of Salvador. The introduction of mechanical transport, such as the automobile, required the old urban structure to adapt to new needs. In the period beginning in 1940, the city's great growth, naturally influencing the center, will lead to a more significant transformation of the landscape.

According to Santos' (1958) reasoning, the center - of Salvador - increasingly had a less central position in relation to the rest of the city. However, even with the growth of the city's outskirts, the center still represented a major junction for all urban circulation. After 8 p.m. and the end of the workers' working day, the center took on a new meaning: there was another type of circulation, that of people coming to seek distraction. Santos (1958) describes:

"It's busiest at the entrance and exit of the movie theaters, which are open until midnight. [...] It's the heart of the city at night, close to the prostitution area, where prostitutes, vagrants and outcasts of all kinds meet in poorly-lit streets. The street vending of fruit and food, cooked or heated under the gaze of the customers in small fires lit on the sidewalks, moves there. Passers-by, still far away, smell the strong aroma of Afro-Brazilian delicacies, seasoned with dende oil and pepper by black and mulatto women dressed in typical costumes. The botequins become busy. The police loosen their vigilance and prostitutes can leave their homes and show off in the street. This is happening in the Upper Town." (page 130)

Reading Santos' account of the urban dynamics of the center of Salvador, observed through its landscape, we could perfectly well be facing the words constructed in Amado's narrative. This passage describes, in fact, the *lumpen* life that Quincas now lived in Pelourinho at the time of his first death. Thus, an urban analysis of a more technical and scientific nature by Santos fits in with our ontological activity on Amado's work. In this passage of Quincas in Amado's book after his second death, we seem to perceive exactly the same geographical environment:

"-Let's give him a drink too... - proposed the Corporal, eager for the dead man's good graces. They opened his mouth and poured the liquor in. It spread a little over the collar of his jacket and the chest of his shirt... They sat Quincas on the box, his head moving to and fro. With the sip of cachaga, his smile widened.[...] Quincas seemed relieved when they took off his black, heavy, very hot jacket. But as he continued to spit out cachaga, they also took off his shirt. Curio looked at the shiny shoes, his own were in disarray. What dead man wants new shoes for, right, Quincas?[...] Quincas leaves the box to eat moqueca from Maria Clara, Master Manuel's wife. Curio and Pe de Vento went ahead. Quincas, satisfied with life, danced between Negro Pastinha and Cabo Martim, arm in arm." (Amado, 1959, pp. 91/92)

The population of the central districts of the city, representing the secondary characters in the narrative, represent their own dynamics of occupation of this place with the noble, old and rich houses that are now also in disrepair. In the meantime, this part of the city is being fought over by other activities, which are gradually driving the population out of certain streets, now occupied by commerce and other activities. The decrease in population corresponds, in certain streets, to the housing crisis and real estate speculation, which generally leads to high rents, unaffordable for poor people, who are forced to build miserable shelters or live in the slums of the center. Quincas' choice: to leave his middle-class home for the city center.

Quincas Berro D'Agua was a resident of a cortigo in Pelourinho. In the center of the city, the cortigos represent the aging palaces and mansions that have lost their former role as residences for nobles and wealthy people, and are now known by other social groups. Some are used exclusively by the poor. Others, on the ground floor, house a trade or handicraft business. The cortigos are "the result of the progressive degradation of these old houses and sobrados, built in the center of the city when this was the wealthy residential part." (Santos, 1958, p. 162).

Pelourinho is the central part of the city where Quincas Berro D'Agua lives. Pelourinho is the location of the tenement he chose to live in. The Pelourinho is an irregularly shaped neighborhood of slopes and slums, surrounded by more vertical buildings, noble two- and three-storey houses that once served as residences for wealthy families and have now fallen into ruin. The description of Pelourinho by Santos (1958) shows us a living landscape, where we are able to construct in this setting all the lumpen and working class population represented in Amado's book:

"The ground floor of all the buildings is occupied by shops and crafts. There you'll find vulcanization workshops, bazaars, tailors, jewelers, houses that buy and sell scrap metal, repairers of various things, warehouses, haberdasheries, cheap restaurants, shoemakers, bakeries, printers, photographers, third-class barbers, bawdry shops, etc.[...] on the floors live a heterogeneous population that lives in more than precarious conditions." (page 171).

Pelourinho, like the rest of this transition zone, is preferred by people who can't afford to pay high rents or spend a lot on transportation. There are many people who don't have a permanent job - this is well portrayed by the Amadian lumpen class - and among the most frequent trades are occupations linked to the informal sector. In short, they are people who have no permanent or defined occupation and whose place of work is preferably in the city center.

Salvador de Santos e Amado is a city in which the informal commercial activities of the center demarcate the various ways of life of the people of that neighborhood. The book *The Death and Dying of Quincas Berro D'Agua is* clearly delimited as having a protagonist and the plot around him, the existential setting of the narrative and that of its own definition. However, the secondary characters around him shape the social dynamics of Salvador's city center even more clearly. Through their habits and their journeys throughout the book.

6 - Quincas' geographical surroundings

With the aim of concluding this last chapter, as well as the final section of the thesis, it was methodologically decided to revisit the concept of the geographical environment and some of its implications. Conceptually, this was a relevant approach in the theoretical development of the thesis as well as a problematization that recurred throughout the text. The biases may have unfolded with each proposed analysis, and a final look at these propositions is now in order. The conceptual importance of the geographical environment has been explored, and examples of it have also been found in the empirical work of Jorge Amado. However, the final proposal is to return to this concept and give it a more ontological than conceptual charge.

One of the aspects that Amado worked on most in his texts was his sensitivity to the subjective issues of the characters. Sensitivity to the geographical environment involves issues such as adapting to reality, collective actions to bring people closer together and even housing needs. Quincas represents a character in which all these situations emerge clearly in his social context and run through his trajectory. Quincas has undergone processes of transformation in his way of life that have reconfigured his reality. The relationships between people throughout the book were formed in the context of their "deaths". Their dwelling references defined not only the geographical environment but also the possible transformations attributed to it. And we will call all this movement around the agendas surrounding Quincas his geographies.

Regarding the concept of the geographical environment, we will finalize its theorization according to the logic of a more ontological nature based on the theorizations of Ruy Moreira (2008). On this concept, he states:

"The geographical environment is a diverse whole of beings, things and men who live together in the same space. Cohabitation is the fundamental aspect since, from a geographical point of view, the fact of cohabitation, that is, the common use of a certain space, is the foundation of everything." (page 67)

With this in mind, let's think about the geographical environment constructed in *The Death and Death of Quincas Wateryell.* Through the sensitivity of adapting to his new reality, the main character - and around him the entire geographical environment was formed - practiced improving his relationship with the environment. The result is his way of life, where a way of structuring his existence shapes and defines him little by little as a being. At the end of the narrative, we wonder if it was Quincas who molded himself to the environment or if a certain geographical environment was born during the construction of his existence.

The layout of the city and its landscape elements, - as analyzed in the previous section - the outline of Salvador's city center with a focus on Pelourinho; the architectural and dilapidated aspect of the houses such as the tenements mentioned as housing; all of this reflects the housing conditions of the Amadian geographic environment. Even more important is to realize that this housing aspect is something that gives identity to the ways of life of human groups and forms their geographies. It is the relationship between man and environment that is constituted and defines the beings of men.

Another important reasoning in the context of the geographical environment is that it is formed in the analysis of this thesis according to urban logic. In other words, we are talking about a geographical environment that is representative of cities. Salvador was analyzed as a city that has experienced processes of expansion and new meanings both in its habitat and in people's way of life.

With the expansion of activities in the city, its symbolic power gains more importance and the city becomes a "geographical entity par excellence сото the means of transportation and communication." (Moreira, 2010, p. 88).

To think about the city in the logic of this thesis was not to realize all the processes of transformation and evolution of the urban landscape and their implications. To understand the city that was problematized about Amado's Salvador was to go through a framework of understanding in which we lend it a sense of meaning, things and relationships of the city becoming ontologically something for Amado's life and its people.

Finally, let's return to the concept of geography. According to Moreira (2010), in order to be geographical, every entity has to be located and situated within a distribution of locations. Location presupposes being in a given place. Being in this sense becomes essential for man's insertion into the world. Geography is at once the being of this man in the world. In this reading, geography is a concept that becomes ontological.

The ontological construction between the geographical environment and the geographies discussed here is one of the most important points of the geographical reasoning in this thesis. With regard to this construction, Moreira (2010) states:

"The relationship between man and the environment, by whatever name we call it, is the ambit of being's struggle to stay alive in order to achieve subsistence. And the living being is the first ontological meaning of geography." (page 157).

Geography is more than a pure context of the geographical environment. It is from there that man builds his sense of existence. And it's important to emphasize that this movement is only possible by being in this geographical environment of the happening of geography. Thus, we return to the dialectical reasoning of the constitution of both.

Returning to Amado and his protagonist Quincas, it is in this plot of experiences that geography appears in the direct form of meanings, written into the character's way of life. It's interesting to reflect on another point at this concluding moment: does geography in *The Death and Dying of Quincas Berro D'Agua* end in the final context of the novel? Doesn't the way it unfolds allow us to see a movement of permanence? And so Jorge Amado would give his work universal status. A final passage in Amado's book that reflects on this idea deals with the fate of Quincas. It is as follows:

"His destiny had been truncated, he who could have become a ship's captain, dressed in a blue uniform, pipe in mouth. Even so, he was still a sailor, and if they gave him that sloop, he would be able to take it across the sea, not to Maragogipe or Cachoeira, nearby, but to the distant shores of Africa, even though he had never sailed." (Amado, 1959, page 54)

At this point in the narrative, Quincas is close to what in the book would be his last death. His last point of birth or finitude of existence. Quincas had been a city dweller, a builder of geographies in the city. However, at this point in his life he has already earned the nickname of sailor, where his idealization would be that of a great sea captain. Even though he had never sailed a sloop, his continuity would be marked by a new sense of location. Was it the definitive death or the beginning of another experience? It's up to everyone's imagination to answer, although both answers define Quincas' being.

Another appropriate reflection for the final moments of the Amadian narrative would be to identify the dialectical construction of the character and the city of Salvador:

"But when the shadows of twilight descended on the city, Quincas became restless. As if he was waiting for something that wasn't long in coming." (Amado, 1959, p. 62)

It is as the city changes its nexus of everyday time, as it prepares to reconfigure the daily routines of its citizens, that Quincas begins to think. Amado takes us back not only to a moment of reflection, but also to a reconstruction of temporalities. In another passage with Quincas' friends, this movement shifts from the protagonist to the secondary characters and becomes clearer:

"The rush had left the five friends, it was as if time belonged to them entirely, as if they were beyond the calendar, and that magical night in Bahia should have lasted at least a week. Because, as Negro Pastinha said, Quincas Berro D'agua's birthday couldn't be celebrated in the short space of a few hours." (Amado, 1959, p. 93/94)

The next and final passage in *The Death and Dying of Quincas Berro D'Agua* will be the chapter's conclusion. It deals with the final moment - in the book - of the protagonist Quincas Berro D'Agua. In this section of the text, some important points need to be noted. Firstly, Quincas' final destination is linked to the geographical environment of Salvador itself, now portrayed as the sea. And, finally, the voice of Quincas defying death and saying that he will leave when he wants to. Quincas' being gains the status of timelessness. Watch then:

"They hoisted the sloop's sails and pulled up the large stone that served as an anchor. The moon had turned the sea into a path of silver, and in the background the black city of Bahia jutted out of the mountains [...] No one knows how Quincas got to his feet, leaning against the smaller sail [...] Stones of sea washed the boat, the wind tried to break the sails. Only the light of Mestre Manuel's pipe persisted, and the figure of Quincas, standing, surrounded by the storm, impassive and majestic, the old sailor. The sloop was approaching the calm waters of the breakwater slowly and with difficulty.[...] Then five bolts of lightning flashed across the sky, the thunder rumbled like the end of the world, a huge wave lifted the sloop.[...] In the midst of the noise, the raging sea, the sloop in danger, they saw Quincas throw himself over and heard his final words:
- I bury myself as I see fit
When you decide
You can keep your box
For a better occasion
I won't let myself be trapped In a shallow grave on the ground And it was impossible to know

O rest of his prayer" (Amado, 1959, p. 103)

The boat's final struggle with the sea seems to be Quincas' destiny: to resist death in order to remain in the world or to leave for good. And the conclusion follows in the belief that there is no ontological meaning - as problematized throughout the text - better synthesized for the end of the chapter than this.

FINAL CONSIDERATIONS

Throughout the course of this research, there have been various reflections and proposed objectives on the interdisciplinary theme of Geography and Literature, in a study of some of Jorge Amado's works. Among the main goals to be achieved in this thesis, our main focus was to be able to think of a specific concept of man in Amado's urban-themed works, which we will work on throughout the text. Who is man in Jorge Amado's work and what are the geographical determinants that surround him were issues that arose and even recurred during this research.

Throughout the creation of this thesis, other elements became relevant to be addressed in the conclusion of this work, as they became important problematizations for the final result. The first question arises from the image of Bahia, specifically Salvador, its capital. What - after all of Amadiana's research into this part of Brazil - is the fundamental mark of Bahia? I conclude that the most appropriate answer would be a mixture. Looking at Bahia through Salvador, the mixture is a social amalgam. Bloods, races, religions, customs, blacks and whites, Indians and Mamelukes, rich and poor, mulattoes and mulattos; this people has emerged in a logic that must be democratic, the condition of all who live there. In this city, which over time has been marked by social inequalities that remain veiled to this day, socio-spatial contradictions have found a way to coexist.

Another important consideration is what kind of view Jorge Amado's work now encompasses and the reading of his personal and political positions throughout his life. I believe that once again the nuance of miscegenation allows for a broader understanding of the writer and his literary proposals. In Amado's case, there is an originality in his work that made Bahians different from other regions of the country, and at the same time, made Bahian man a being of spatial practices derived from a geographical environment that was ontologically possible to define. This is a hybrid and diverse geographical environment, further confirming the mestigagem content of the place, and precisely this proposition explains its genuine formation.

It has been written more than once that an artist - writer, artist, musician - will be all the more universal when he or she is national and that his or her work will gain immortality to the extent that it bears the mark of its time, bears witness to it and participates in it. And, in the case of Amado, it has become quite clear over the years of research that his work - the most widely read Brazilian regional literature outside the country - has risen to universal status precisely because it tries to maintain the characteristics of its regionalist content. Jorge Amado is the reader from the Northeast of Bahia who has brought the most visibility to this part of Brazil. This originality, this national/regional face, the color of our nation formation and our skin, our physiognomy, cannot be forgotten for a single minute once it intends to assert itself as an entity and a being. Jorge Amado brought the construction of a man typical of his surroundings throughout his narratives, because he created by meeting the spectacle of his people, absorbing the prevailing reality, making the temporal world an eternal existence.

Jorge Amado's originality lies in giving birth to an urban regionalism that until then had not been worked on in the Northeast when he began writing his first works. Amado's regionalism is that of the nascent proletariat, which found in our author both its expression and its embodiment. This is how Amado is seen, because he was a novelist who only fantasized what he lived. Amado doesn't write with an external eye to the content of his work, he writes about what he experienced. The men of the people, blacks, mulattos, street kids, "bitches", women of struggle, were Jorge Amado's real teachers.

Jorge Amado was a staunch supporter of communist ideas and a member of revolutionary movements. He was exiled from his country from 1948 and this period was also very expressive in his works from then on. Prison and exile forced Jorge Amado to distance himself from his country and rebuild it through imagination, memory and nostalgia, thus giving it that dimension of universality that makes his heroic brothers understandable to people from all countries and all races, whatever their skin color. Jorge Amado's art consisted - as evidenced by the success of his novels, which have been translated into countless languages - in transforming a well-characterized regional category, the Brazilian Northeast of Bahia, into a universal category.

Another important point to be raised at this final stage of the thesis is why Jorge Amado was chosen as the basis for an ontological reading of geography. There were several moments when the answer didn't seem to be clear and concrete during the intertwining and difficulties in the scientific development of the thesis. Jorge Amado did not remain linear in his defense of political ideas and thoughts throughout his life and fruitful literary production. He changed over time and reformulated some of his convictions, such as his unquestioning defense of the Communist Party and Soviet socialism, which was no longer the case at a certain point in his life. However, even with the different approaches his novels took over time in terms of theme (culture, raga, religion, feminist protagonism, humour, protest) or place (countryside, city, southern Bahia, Salvador), one point characterizes the unity of all his literary production. It is precisely this that has allowed this ontological reading to reach the geography of this thesis. Apart from all these changes in style and way of writing, what makes Jorge Amado's thought unified is his defense of *Being* against *Having*. The defense of the spontaneity of life against the illusory pursuit of material wealth or the appearances of respectability, of freedom, in short, against forms of self-alienation that are well understood as forms of oppression.

Marxism was the means by which our writer was able to give his painting a universal character, making the case of the Bahian landowner the particular example of a much more general phenomenon, that of the exploitation of man by man. Jorge Amado's Marxism was sometimes quite pamphleteering and at other times it was implicit in the inner psychological characterizations of his characters. Thus, his Marxism was much more than a political ideology, but an artistic procedure. The process by which humor detaches itself from the singular in order to reach the universal. In this Marxism that founds his characters, we see the birth of the Amadian man of Salvador. We see Salvador as a city full of social contradictions, since it is defined by the realization of the geographies of this very man, this social being. Amado's Marxism is also messianic, almost religious; he looks more to the paradise of the future than to the interpretation of the past: tomorrow will be better and more beautiful.

The universal nexus in Jorge Amado's work does not remain a simple abstract or disembodied notion of geographical environment, or even of meaning. This universal, rooted in Bahian culture, is what has allowed us to read the geographical foundation of the man we seek throughout this thesis. The universal, on the contrary, is consistent in relationships, in a certain culture, both African and European; in a certain geographical environment, where the sea dialogues with the city in movement; in a certain social environment, that of the struggle for the right to the city; in a contrast of coexistence between a petty bourgeoisie in the cult of appearances and vagabonds in search of freedom.

The geography of Jorge Amado's characters lies in the vividness with which they are defined. Amado's characters are as alive as the people who live in the city of Salvador today. They walk, cry, get drunk, make love, kill and are killed. He tells of the actions of concrete beings, immersed in reality, sometimes even in symbiosis with their surroundings, such as the sea, the hills and the city itself.

In Amado's narratives, the sea, the night and the storm are more than just themes; they are characters, as real, as alive, as Balduino, Pedro Arcanjo, Livia, Guma, Quincas or Jubiaba. Thus, we glimpse this setting in Jorge Amado's plots, the foundation of the geographical environment and ontologically important precisely because it is what makes the dialectic existence between man and environment real. In *Dead Sea,* the sea plays this role. In *Tent of Miracles, it's* the tent itself that gives the book its title. In Jubiaba, Antonio Balduino's protagonist constructs his geographies in his journeys, but there is one that stands out: Morro do Capa Negro. This is where we see his before and after existence. And, in Quincas, Pelourinho gains this status. Thanks to these aspects, Jorge Amado inscribed his work in the world, also giving it a teluric sense.

In concrete terms, Bahia is both misery and poetry, misery ending in poetry and poetry translating man's reaction to misery, concretely personified in Amado's living and present-day characters. He kept in the creative mind of his narratives the image of streets, market stalls, churches, the port pier, xangos and candombles; of black samba circles, because he was a miniature of Bahia and the Northeast. Laughter and the harshness of inequality coexist as allies in Jorge Amado's work. They liberate all self-alienation and all oppression, making it possible to destroy social ties and find the fluidity of life.

At this point, I believe that the meaning of man in Jorge Amado has already been answered by all the content worked on in the thesis. However, in confirmation, the man in Jorge Amado is the one who moves ontologically from the collective vision to the individual vision. He is the being who constructs his geography in the spatial cut-out of the world that is Salvador. He is the being who also makes his geographic foundation out of the scenery of circulation through which he lives in the city of Salvador. The internal temporalities of Amadian beings are conceived in two different ways: the intensity of sensation generates the 'instant'; the multiplicity of sensations produces 'duration'. Therefore, the supreme venture, plenitude, would be in having strong and continuous sensations. There's no shortage of this in Jorge Amado's realistic stories, just as there are plenty of these conjunctures.

The importance of Bahia's urban setting in the works was not only a strategic focus for the thesis, but also due to the belief that the city is at the heart of the current movement of social problems. Brazil has become urban and so has Bahia. Salvador was the country's first capital and represents one of the oldest cities in the construction of national identity. The city appears in typical form in his works, the suburban city and the picturesque outskirts. Jorge Amado is also an urban writer because he talks about his people in their own language. The base is the people, the speech is of the people. The identification of the language of the people emerges as an existential ground. The writer is surrounded by his characters, identified with them, listening to them at every moment in this urban landscape in which they live.

Jorge Amado's city is not just a city-habitat. It is a city that has habits, that has people and not just numbers. It is the city that forms a man. The perception of the city is something instigating and this perception can occur as a written scene of the city that remains. This is what happened with the geographical reading of Jorge Amado's works. It is therefore possible to decipher the city by reading its history told in the novels.

The general aim of this thesis was to analyze the geographical foundation of Jorge Amado's works from an ontological point of view, where the definition of man would be based on his geography. This man can be diverse and post-modern, given the questions about the past that Amado posed in his works. However, the timeless vision of his work has not been lost, and still allows us to place his particularities on a universal scale of existence. The Amado man is the man from Bahia, the regional man from the Northeast and also a man of the world.

BIBLIOGRAPHICAL REFERENCES:

ALBUQUERQUE JUNIOR, Durval Muniz de. "A invengao do Nordeste e outras artes". Ed. Cortez, Sao Paulo, 2001.

ANDRADE, Adriano Bittencourt. "The city of Salvador in 1959: the views of Jorge Amado and Milton Santos". IN: Visoes imaginarias da cidade da Bahia. Institute of Geosciences, UFBA, Salvador, 2004.

AMADO, Jorge. "The death and death of Quincas Berro D'Agua". Ed. Record, Rio de Janeiro, 1959.

AMADO, Jorge. "Caderno de Literatura", Institute Moreira Salles, Rio de Janeiro 1997.

AMADO, Jorge. "Capitaes da Areia" Ed. Martins, Sao Paulo, 1937.

AMADO, Jorge. "Jubiaba", Jose Olympio, Rio de Janeiro, 1935.

AMADO, Jorge. "Dead Sea", Jose Olympio, Rio de Janeiro, 1936.

AMADO, Jorge. "Tent of Miracles", Ed. Martins, Sao Paulo, 1969.

AMADO, Jorge. "Terras do sem firn": Ed. RECORD, Rio de Janeiro, 1985.

BAKHTIN, Mikhail. "Marxism and Philosophy of Language". Ed. Hucitec, Sao Paulo, 1981.

BASTIDE, Roger. "Narrative is a socially symbolic act". Ed. Atica" IN: *Jorge Amado Povo e Terra - 40 years of Literature.* Sao Paulo, Ed. Martins, 1972.

BERGAMO, Edvaldo. "Fiction and conviction: Jorge Amado and Portuguese literary neo-realism", Ed. Unesp, Sao Paulo, 2008.

BESSE, Jean-Marc. "Geography and existence: from the work of Eric Dardel". IN: *Man and the earth; the nature of geographical reality.* Ed. Perspectiva, Sao Paulo, 2011. BESSE, Jean-Marc: "Vera Terra", Ed. Perspectiva, Sao Paulo, 2006.

BLANC, *Mafalda* de Faria. "*Introduction* to *Ontology".* Instituto Piaget, Lisbon, 1997.

BOSI, Alfredo. "Historia Concisa da Literatura Brasileira", Ed. Cultrix, Sao Paulo, 2006. Candido, Antonio. "Iniciagao a Literatura brasileira: resumo para phneipiantes", Edusp, Sao Paulo, 1999.

CANDIDO, Antonio. "Literature and Society". Ed. Ouro sobre Azul, Rio de Janeiro, 2006. CARNEIRO, Marlene Pires D'Aragao. "0 homem e o lugar - Pelourinho um olhar no tempo e no espago social de Quincas Berro D'Agua" IN: *Visoes imaginarias da cidade da Bahia: um diálogo entre a geografia e a literatura.* Edufba, Salvador, 2004.

CAUQUELIN, Anne. "The Invention of Landscape". Ed. Martins Fontes, Sao Paulo, SP, 2007. DA MATTA, Roberto. "0 que e o Brasil", Ed. Rocco, Rio de Janeiro, 2003.

DARDEL, Eric. "0 homem e a terra: natureza da realidade geografica". Ed. Perspectiva, Sao Paulo, 2011.

DUARTE, Eduardo de Assis. "Jorge Amado: Romance in a time of utopia". Ed: Edufrn, Rio Grande do Norte, 1995.

FARACO; Carlos Alberto, CASTRO, Gilberto de; TEZZA, Cristovao (eds). "Dialogues with Bakhtin", Ed. UFPR, Curltlba, 2001.

FOUCAULT, M. Les mots et les choses. Paris: Gallimard, 1966.

FREYRE, Gilberto. "Casa Grande & Senzala". Portugal. 1933.

FREYRE Gilberto. "Sobrados e Mocambos". 1936.

FRIEDMANN, Georges. "7 studies on man and technology", Ed. Difusao europeia do livro, Sao Paulo, 1968.

GILES, Thomas Ranson. "Jean Paul Sartre" In: Historia do Existencialismo e da Fenomenologia, Edusp, Sao Paulo, 1975.

GOBINEAU, Arthur de. "The moral and intellectual diversity of the ragas", 1855.

HOLZER, Wherter. "The phenomenological geography of Eric Dardel". IN: *Man and the earth: the nature of geographical reality.* Ed. Perspectiva, Sao Paulo, 2011.

INWOOD, Michael. "Heidegger", Edigoes Loyola, Sao Paulo, 1997.

JAMESON, Fredric. "Political unconscious: the narrative сото socially symbolic act". Ed. Atica, Sao Paulo, 1992.

JAMESON, F. "Postmodernism and Politics". Ed. Rocco, Rio de Janeiro, 1992.

JAMESON, Fredric. "Is the historical novel still possible?", Translated by Hugo Mader. Novos estudos CEBRAP, Sao Paulo, n° 77, 185-203, 2007.

LESSA, Sergio. "Lukacs and ontology: an introduction", Revista Outubro, Ed. 05, Alagoas, 1983

LLOSA, Vargas. IN: Caderno de Literatura: Jorge Amado, Instituto Moreira Salles, Rio de Janeiro, 1997.

LUCAS, Fabio. "Plano, com Epfgrafe, de um estudo sobre a Morte de Quincas Berro D'Agua", IN: *Jorge Amado Povo e Terra - 40 anos de Literatura,* Ed. Martins, Sao Paulo, 1972.

LUKACS, Georg. "The Ontological Bases of Man's Thought and Activity", Revista Temas de Ciencias Humanas. Sao Paulo, v. 1, p. 1-18, 1979.

LUKACS, Georg. "Introduction to Marxist Aesthetics". Civilizagao Brasileira, Rio de Janeiro, 1970.

LUKACS, Georg. "Prolegomenos para uma ontologia do ser social". Boitempo, Rio de Janeiro, 2010.

LUKACS, Georg. "The Theory of the Novel", Editora 34, 2ed., Sao Paulo, 2009.

MACHADO, Ana Maria. "Romantic, Seductive and Anarchist: сото and why read Jorge Amado today". Objetiva, Rio de Janeiro, 2006.

MARANDOLA JR, Eduardo and GRATAO, Lucia Helena Batista. "Geography and Literature: essays on geography, poetry and imagination". (eds). Eduel, Londrina, 2010.

MARTINS, Elvio Rodrigues. "Geography and Ontology: the geographical foundation of being" IN: Geousp, n° 21, Sao Paulo, 2007.

MARTINS, E. R. "Pensamento Geografico e Geografia em Pensamento". In: Angela Massumi Katuta, Deise F. Ely, Eliane T. Paulino, Fabio C.A. da Cunha, Ideni T. Antonello. (Org.). Geografia e Midia Impressas. led.Londrina: Universidade Estadual de Londrina, 2009, v., p. 13-36.

MONTEIRO, Carlos Augusto de Figueiredo. "O mapa e a trama: ensaios sobre o conteúdo geografico em criagoes romanescas". Editora da UFSC, Florianopolis, 2002.

MOREIRA, Ruy. "The forms of Geography and the work of the geographer over time" IN: Thinking and

Being in Geography. Contexto, Sao Paulo, 2010.
MOREIRA, Ruy. "Geography serves to unveil social masks" IN: Thinking and Being in Geography, Contexto, Sao Paulo, 2010.
MOREIRA, Ruy. "The spatial categories of the geographical construction of societies". IN: Thinking and Being in Geography. Contexto, Sao Paulo, 2010.
MOREIRA, Ruy. "Concepts, categories and logical principles for the method and teaching of Geography". IN: Pensare Serem Geografia. Contexto, Sao Paulo, 2010.
MOREIRA, Ruy. "Identity and the representation of difference in Geography". IN: Thinking and Being in Geography. Contexto, Sao Paulo, 2010.
MOREIRA, Ruy. "Sociability and space: societies in the era of the Third Industrial Revolution". IN: Pensare Serem Geografia. Contexto, Sao Paulo, 2010.
MOREIRA, Ruy. "Ser-tao: o Universal no Regionalismo de Graciliano Ramos, Mario de Andrade e Guimaraes Rosa (um ensaio sobre a geograficidade). IN: *Ciencia Geografica,* vol. X, Bauru, 2004.
OLINTO, Antonio. *"Tenda dos Milagres: magic and revolution in Portuguese-language literature".* Ed. Martins, Sao Paulo, 1972.
PORTELLA, Eduardo. "A fabula em cinco tempos". IN: *Jorge Amado Povo e Terra - 40 anos de Literatura.* Ed. Martins, Sao Paulo, 1972.
PRANDI, Reginaldo. "Mythology of the Orishas", Ed. Companhia das Letras, Sao Paulo, 2001.
RAILLARD, Alice. "Conversing with Jorge Amado". Ed. Record, Sao Paulo, 1990.
RAMOS, Artur. "O Folk-lore Negro do Brasil: demopsicologia e psicanalise" Casa do Estudante do Brasil, Rio de Janeiro, 1935.
RIBEIRO, Darci. IN: "Caderno de Literature: Jorge Amado", Institute Moreira Salles, Rio de Janeiro, 1997.
RODRIGUES, Nina. Africans in Brazil. 4 ed. Cia. Editora Nacional, Sao Paulo, 1976.
ROMERO, Silvio."O Brazil Social". in Revista do Inst. Historico e Geografico Brazileiro, tomo LXIX, Parte II, Rio de Janeiro, 1908.
ROSSI, Luiz Gustavo Freitas. "As cores da Revolugao: a literatura de Jorge Amado nos anos 30". Ed.Unicamp, Sao Paulo, 2009.
SANTOS, Douglas. "A reinvengao do espago", Editora Unesp, Sao Paulo, 2002.
SANTOS, Milton. "The nature of space". Hucitec, Sao Paulo, 1988.
SANTOS, Milton "O centro da cidade de Salvador", Salvador, 1959.
SELLMAN, Mauricio. "The construction of a city: Salvador in the writings of Jorge Amado", PhD thesis, Manchester, 2013.
SCHWARCZ, Lilia. "0 espetaculo das ragas: cientistas, instituigoes e pensamento racial no Brasil 1870-1930". Companhia das letras, Sao Paulo, 1993.
SILVA, Armando Correa da. "0 espago fora do lugar". Hucitec, Sao Paulo, 1978.
SILVA, Armando Correa da. "As categorias coтo fundamentos do conhecimento geografico". IN: 0

espago interdisciplinar. Nobel, Sao Paulo, 1986.

SOJA, Edward W. "Postmodern geographies: the reaffirmation of space in critical social theory". Ed. Zahar, Rio de Janeiro, RJ, 1993.

VAZ, Henrique C. de Lima. "Anthropologia Filosofica". Volumes I and II. Loyola, Sao Paulo, 1992.

VIANNA, Oliveira Francisco Jose Oliveira *"Ra$a e assimilaqao".* 2ª ed, Companhia Editora Nacional. Brazilian Pedagogical Library. Series V. Brasiliana. Sao Paulo, 1934

Printed by Books on Demand GmbH, Norderstedt / Germany